Holt Mathematics

Chapter 12 Resource Book

HOLT, RINEHART AND WINSTON
A Harcourt Education Company
Orlando • **Austin** • New York • San Diego • London

Printed in the United States of America

ISBN 0-03-078403-4

6 7 8 170 09 08

CONTENTS

Blackline Masters

Date ___________

Dear Family,

In this chapter, your child will learn how to identify and graph linear equations and about linear relationships in algebra. Linear equations are used in life science, space travel, manufacturing, and economics.

A **linear equation** is an equation whose solutions fall on a straight line.

If an equation is linear, a constant change in the x-value corresponds to a constant change in the y-value. The graph shows an example where each time the x-value increases by 3, the y-value increases by 2.

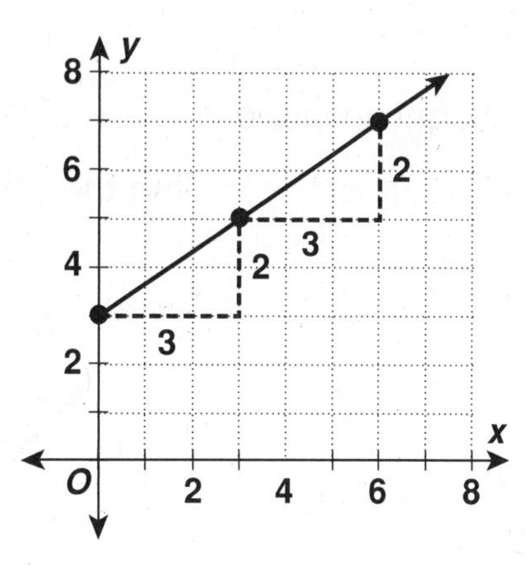

Your child will learn to graph equations.

Graph the equation $y = 2x + 1$.

x	$2x + 1$	y
2	2(2) + 1	5
1	2(1) + 1	3
0	2(0) + 1	1
−1	2(−1) + 1	−1
−2	2(−2) + 1	−3

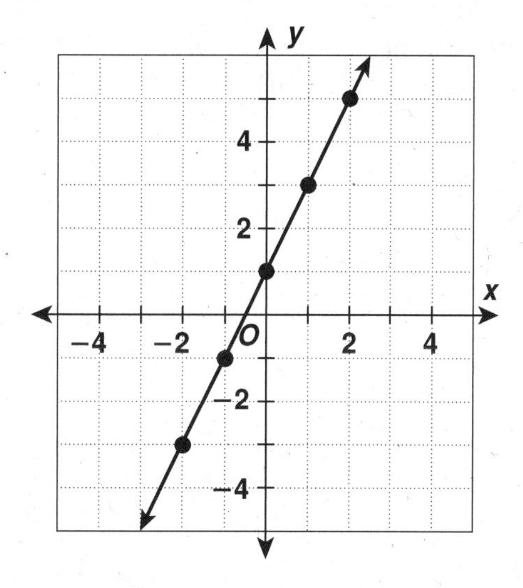

The equation $y = 2x + 1$ is a linear equation because its graph is a straight line and each time x increases by 1 unit, y increases by 2 units.

Slope defines the "slant" of a line. The larger the absolute value of the slope, the steeper or more vertical the line will be.

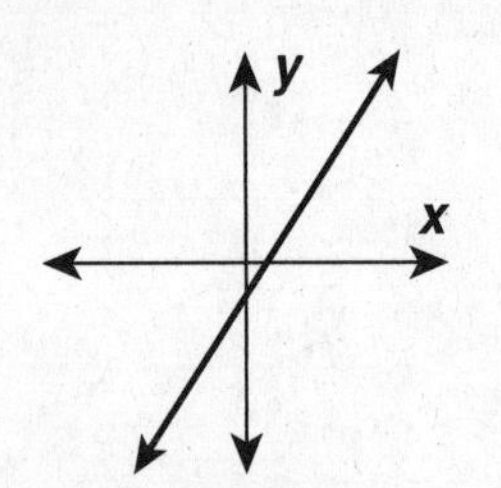

Positive slope

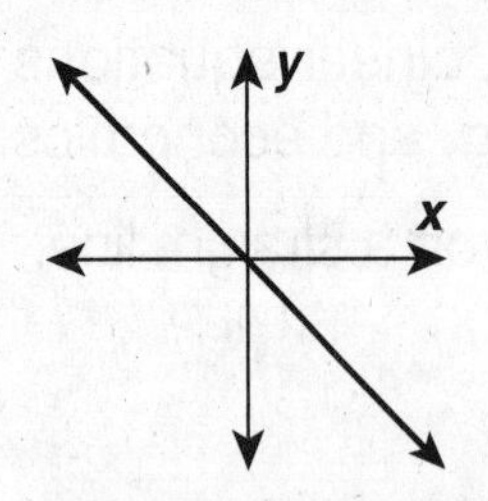

Negative slope

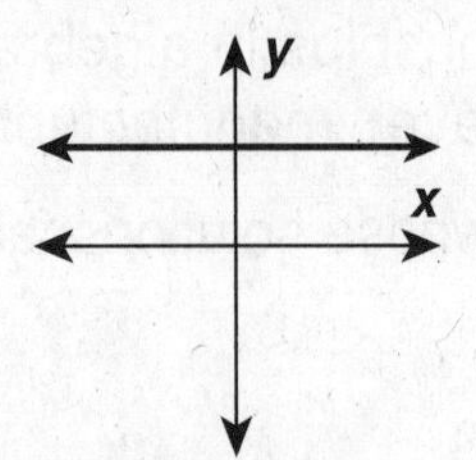

Zero slope

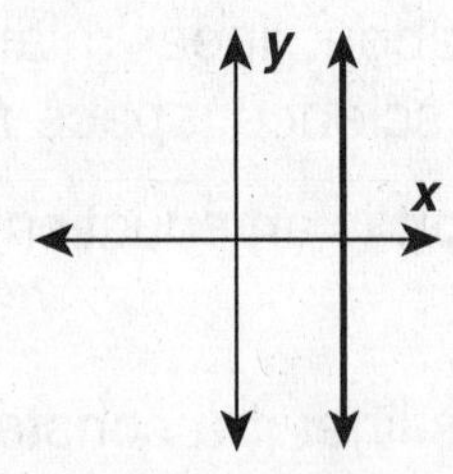

Undefined slope

You can graph a linear equation by finding the ***x*-intercept** and ***y*-intercept**.

The *x*-intercept of a line is the *x*-coordinate of the point where the line crosses the *x*-axis (where $y = 0$).

The *y*-intercept of a line is the *y*-coordinate of the point where the line crosses the *y*-axis (where $x = 0$).

Graph of the line $y = -x + 2$

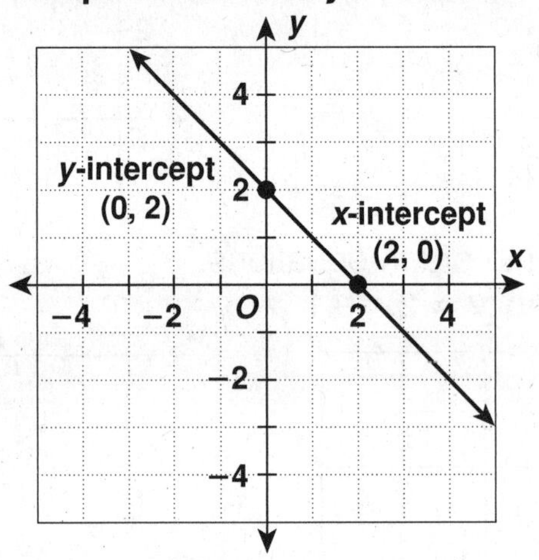

After finding the *x*-intercept and *y*-intercept of the line $2x + 3y = 6$, students can graph the equation.

Find the *x*-intercept ($y = 0$).

$$2x + 3y = 6$$
$$2x + 3(0) = 6$$
$$\frac{2x}{2} = \frac{6}{2}$$
$$x = 3$$

The *x*-intercept is 3.

Find the *y*-intercept ($x = 0$).

$$2x + 3y = 6$$
$$2(0) + 3y = 6$$
$$\frac{3y}{3} = \frac{6}{3}$$
$$y = 2$$

The *y*-intercept is 2.

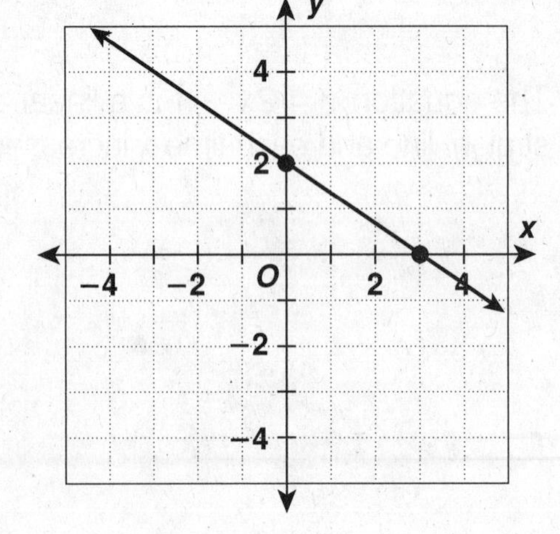

For additional resources, visit go.hrw.com and enter the keyword MT7 Parent.

Name ______________________ Date ____________ Class ____________

LESSON **12-1**

Practice A

Graphing Linear Equations

Complete the tables for the given equations.

1. $y = 3x + 2$

x	$3x + 2$	y	(x, y)
−3	3(−3) + 2	−7	(−3, −7)
−2			
−1			
0			
1			
2			
3			

2. $y = -x - 3$

x	$-x - 3$	y	(x, y)
−3			
−2			
−1			
0			
1			
2			
3			

Complete the table for the given equation. Then graph the equation.

3. $2x - y = 1$

x	$2x - 1$	y	(x, y)
−3			
−2			
−1			
0			
1			
2			
3			

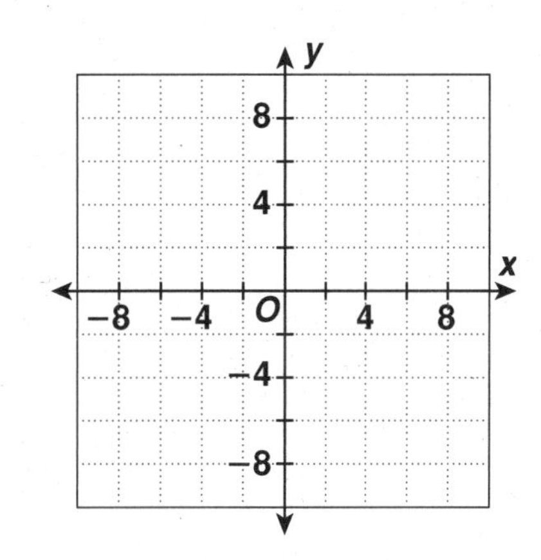

4. Is the equation in Exercise 3 linear? Explain.

____

____

Name ______________________ Date ___________ Class ___________

LESSON 12-1

Practice B

Graphing Linear Equations

Graph each equation and tell whether it is linear.

1. $y = -2x - 5$

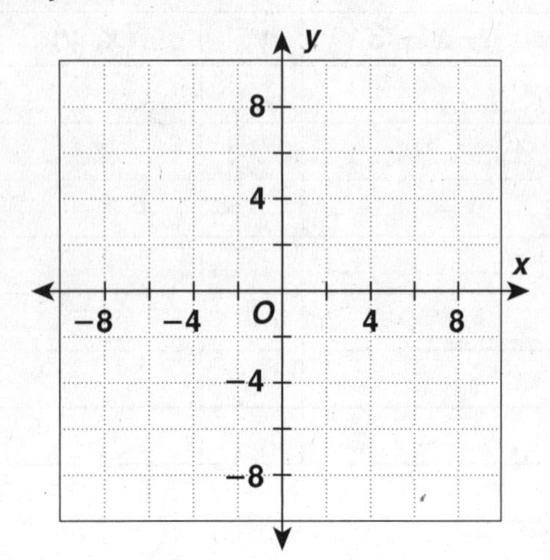

____

2. $y = -x^2 + 1$

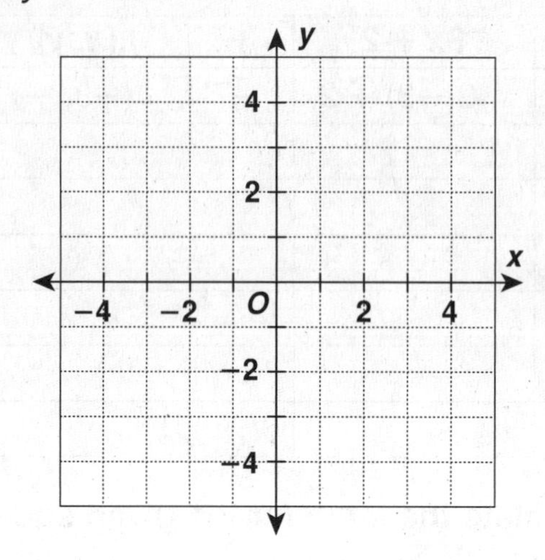

____

3. $y = x^2 - 7$

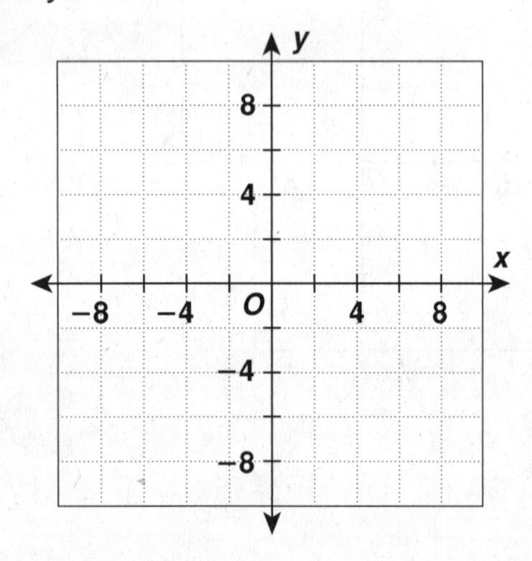

____

4. $y = \frac{1}{2}x - 1$

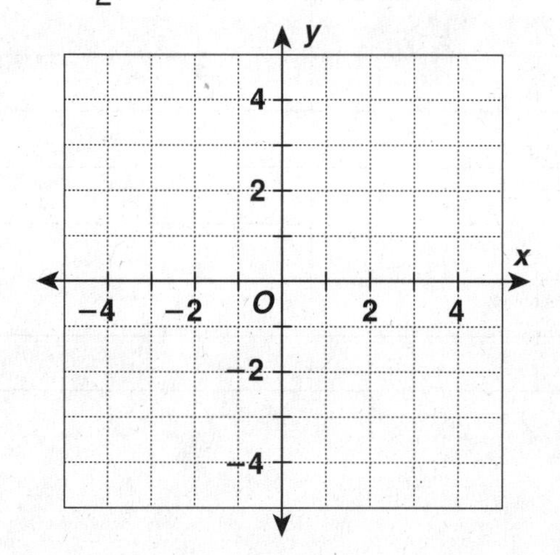

____

5. A real estate agent commission may be based on the equation $C = 0.06s + 450$, where s represents the total sales. If the agent sells a property for $125,000, what is the commission earned by the agent? Graph the equation and tell whether it is linear.

____

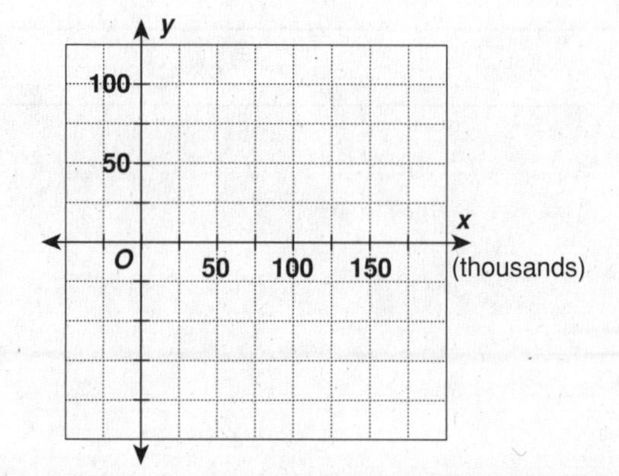

Name ______________________ Date ____________ Class ____________

LESSON **12-1**

Practice C

Graphing Linear Equations

Graph each equation and tell whether it is linear.

1. $y = -3x$

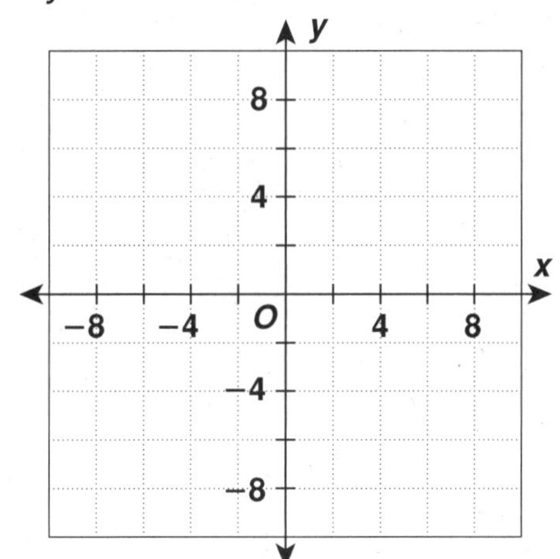

2. $y = x - 5$

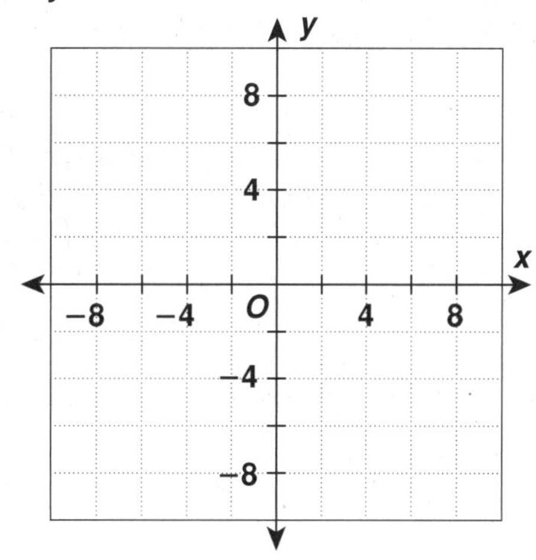

3. $y = \frac{1}{2}x - 3$

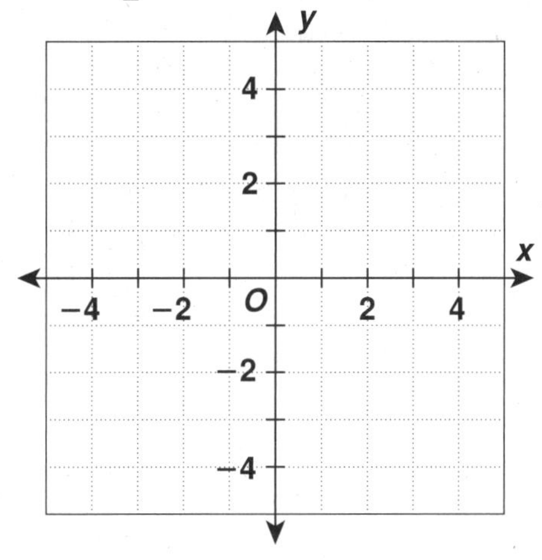

4. $y = -x^2 - 2$

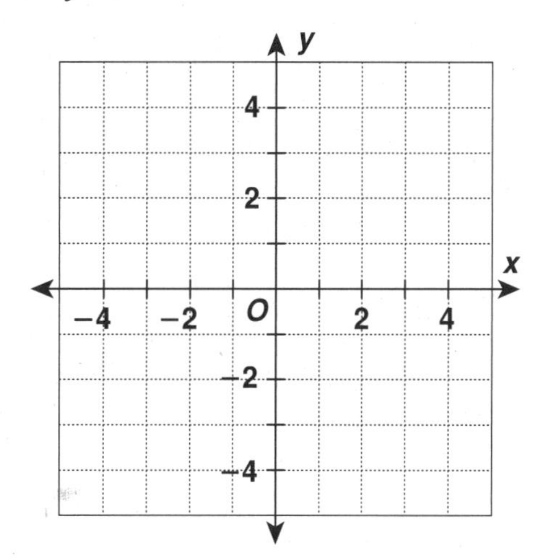

5. Mrs. Blanche grades a math test with a curve based on the formula $G = 5.5p + 10$, where G is the curved grade and p represents the number of problems correct. If Sebastian had 10 problems correct, Alisha had 12 problems correct, and Miguel had 14 problems correct, what was each student's curved grade? Graph the equation and tell whether it is linear.

____

____

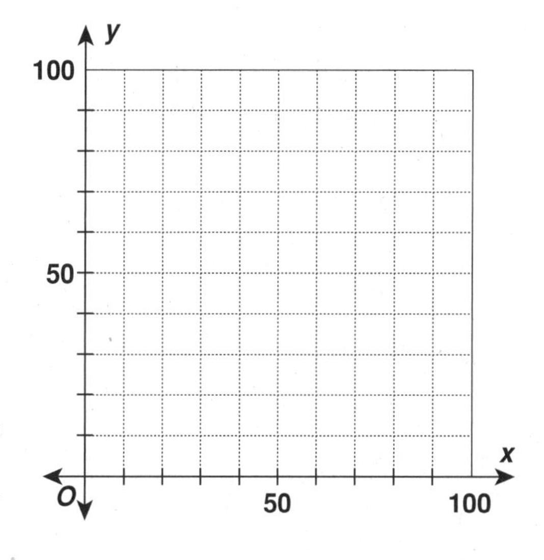

Name ______________________ Date ____________ Class ____________

LESSON **12-1**

Reteach

Graphing Linear Equations

The graph of a **linear equation** is a straight line.

The line shown is the graph of $y = \frac{3}{2}x + 1$.

All the points on the line are solutions of the equation.

Each time the x-value increases by 2, the y-value increases by 3. So, a constant change in the x-value corresponds to a constant change in the y-values.

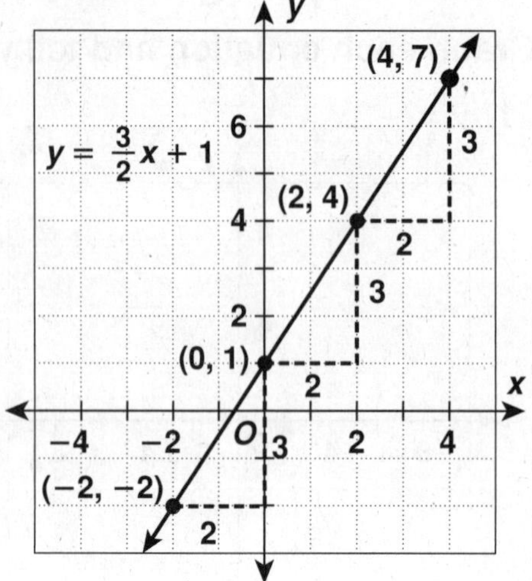

$y = 3x - 4$

Changes in x: +1, +1, +1, +1

x	−2	−1	0	1	2
y	−10	−7	−4	−1	2

Changes in y: +3, +3, +3, +3

Since a constant change in the x-value corresponds to a constant change in the y-value, $y = 3x - 4$ is a linear equation.

$y = 3x^2$

Changes in x: +1, +1, +1, +1

x	−2	−1	0	1	2
y	12	3	0	3	12

Changes in y: −9, −3, +3, +9

Since a constant change in the x-value does not correspond to a constant change in the y-value, $y = 3x^2$ is not a linear equation.

Each equation has a table of solutions. Indicate the changes in x-values and in y-values. Tell whether the equation is linear.

1. $y = 2x - 5$

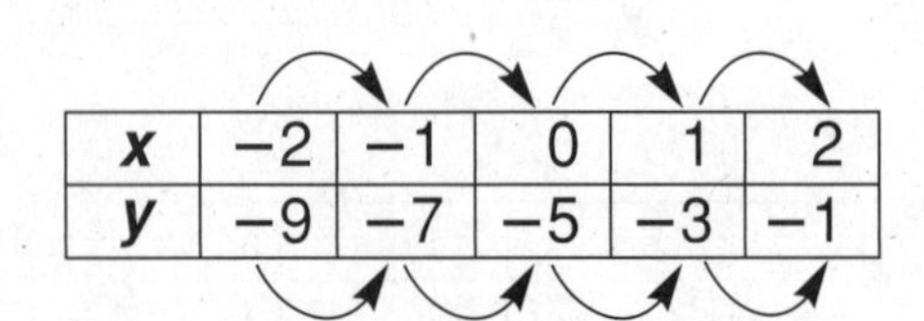

x	−2	−1	0	1	2
y	−9	−7	−5	−3	−1

____

2. $y = 2x^3$

x	−2	−1	0	1	2
y	−16	−2	0	2	16

____

Name ________________________ Date ____________ Class ____________

LESSON 12-1

Reteach

Graphing Linear Equations (continued)

To graph a linear equation, make a table to find several solutions. Choose *x*-values that are easy to graph. Substitute each *x*-value into the equation to find the corresponding *y*-value. Plot your solutions and draw a line connecting them.

x	$\frac{1}{2}x - 3$	y	(x, y)
−4	$\frac{1}{2}(-4) - 3$	−5	(−4, −5)
−2	$\frac{1}{2}(-2) - 3$	−4	(−2, −4)
0	$\frac{1}{2}(0) - 3$	−3	(0, −3)
6	$\frac{1}{2}(6) - 3$	0	(6, 0)

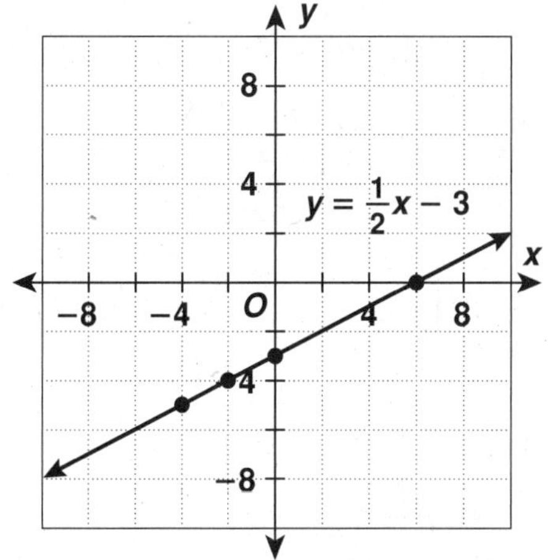

Complete the table for each equation and then graph the equation.

3. $y = 3x - 1$

x	$3x - 1$	y	(x, y)
−2	3(−2) − 1		(−2,)
0	3( ) − 1		(0,)
1	3( ) − 1		(1,)
3	3( ) − 1		(3,)

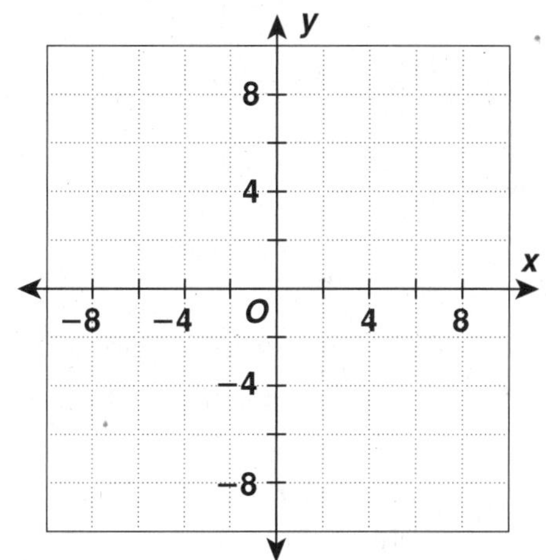

4. $y = -\frac{3}{2}x + 1$

x	$-\frac{3}{2}x + 1$	y	(x, y)
−2	$-\frac{3}{2}(\quad) + 1$		(−2, 4)
0	$-\frac{3}{2}(\ ) + 1$		
2	$-\frac{3}{2}(\ ) + 1$		
4	$-\frac{3}{2}(\ ) + 1$		

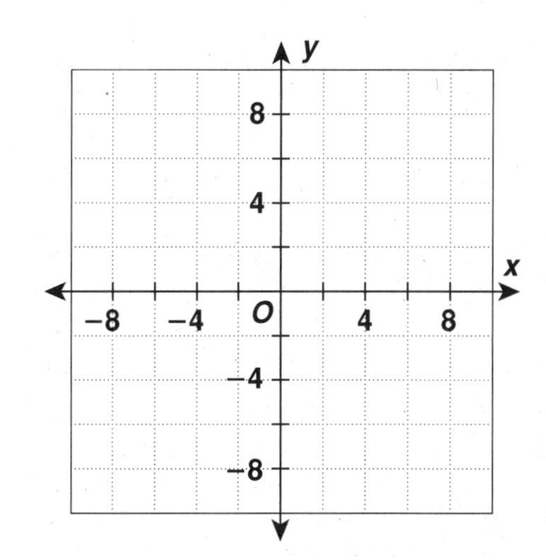

Name ______________________ Date __________ Class __________

LESSON 12-1

Challenge

A Recognition Factor

Different kinds of equations have different kinds of graphs. By studying the graphs of different kinds of equations, you can learn to recognize characteristics of the equations.

1. Complete the table of values to graph each equation. Draw all the graphs on the given grid. Write each equation near its graph.

a. $y = 2x + 1$

x	y
−4	
−3	
−2	
−1	

b. $y = x^2 + 1$

x	y
−1	
0	
1	
2	

c. $xy = 6$

x	y
1	
2	
3	
6	

d. $x + y = -1$

x	y
−5	
−4	
−3	
−2	

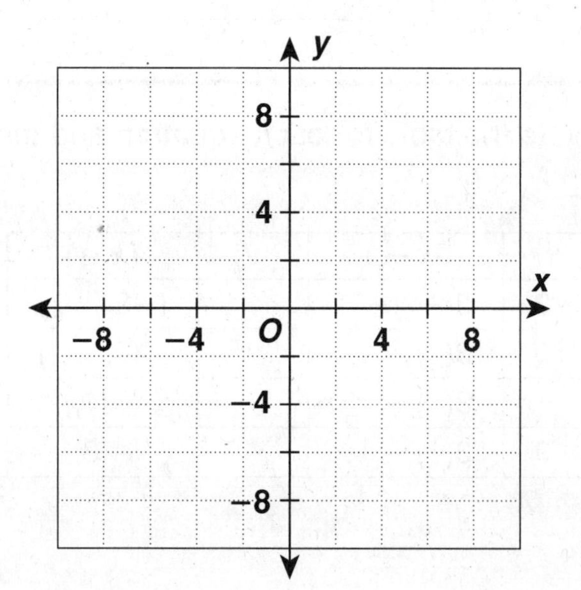

2. Analyze the equations in Exercise 1 and your graphs of the equations. Make a conjecture about how you might recognize a linear equation without graphing it.

____

____

Name ____________________ Date __________ Class __________

LESSON 12-1

Problem Solving

Graphing Linear Equations

Write the correct answer.

1. The distance in feet traveled by a falling object is found by the formula $d = 16t^2$ where d is the distance in feet and t is the time in seconds. Graph the equation. Is the equation linear?

____

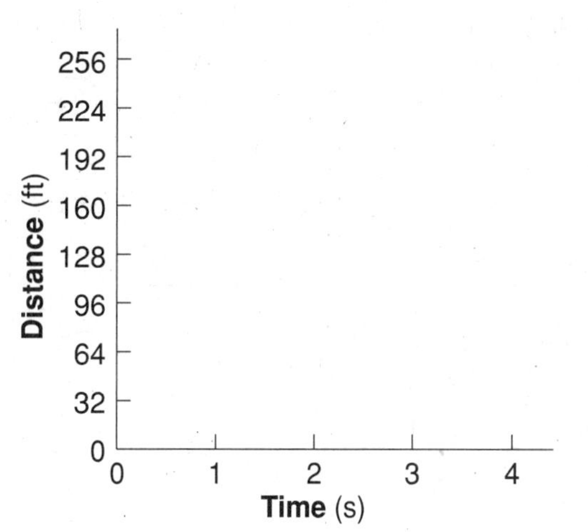

2. The formula that relates Celsius to Fahrenheit is $F = \frac{9}{5}C + 32$. Graph the equation. Is the equation linear?

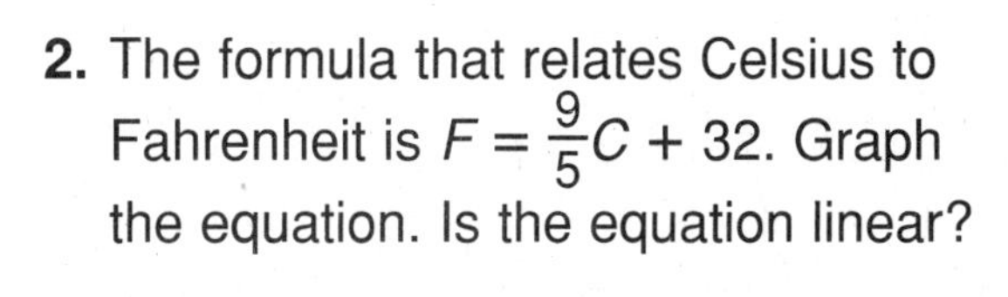

____

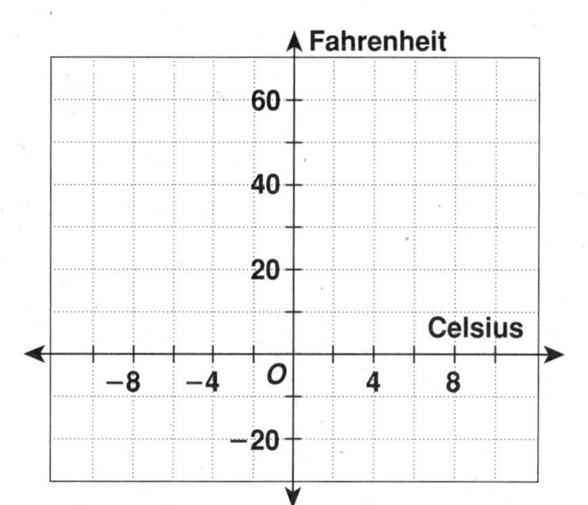

Wind chill is the temperature that the air feels like with the effect of the wind. The graph below shows the wind chill equation for a wind speed of 25 mph. For Exercises 3–6, refer to the graph.

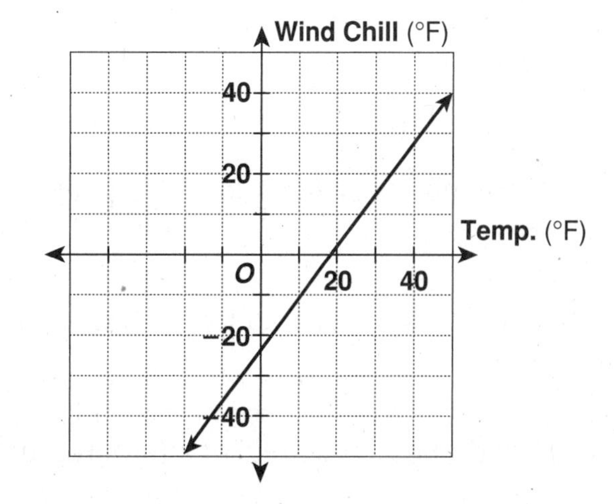

3. If the temperature is 40° with a 25 mph wind, what is the wind chill?

A 6° **C** 29°

B 20° **D** 40°

4. If the temperature is 20° with a 25 mph wind, what is the wind chill?

F 3° **H** 13°

G 10° **J** 20°

5. If the temperature is 0° with a 25 mph wind, what is the wind chill?

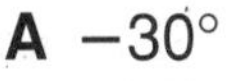

A −30° **C** −15°

B −24° **D** 0°

6. If the wind chill is 10° and there is a 25 mph wind, what is the actual temperature?

F −11° **H** 15°

G 0° **J** 25°

Name ______________________ Date ____________ Class ____________

LESSON 12-1

Reading Strategies
Use a Graphic Organizer

This graphic organizer can help you understand linear equations.

Definition

A **linear equation** is an equation with solutions that lie in a straight line when graphed.

Linear Equations

Table of ordered pairs for $y = 3x - 5$:

x	y	(x, y)
2	1	(2, 1)
3	4	(3, 4)
4	7	(4, 7)

Graph of ordered pairs for $y = 3x - 5$:

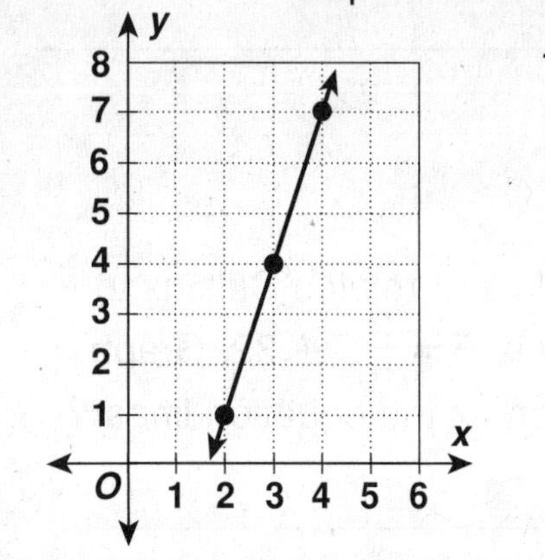

Use the information in the graphic organizer to answer the following questions.

1. What is a linear equation?

____

2. Write the linear equation shown in the example above.

___ = ___ − ___

3. How many solutions for the equation are plotted on the graph?

____

4. How is each ordered pair shown on the graph?

____

5. Write one of the solutions (ordered pairs) shown on the graph.

____

6. What figure is formed when the points on the graph are connected?

____

Name ______________________ Date __________ Class __________

LESSON 12-1

Puzzles, Twisters & Teasers

Straight and Narrow?

Decide whether each graph is linear or nonlinear. Circle the letter above your answer. Use the letters to solve the riddle.

1.

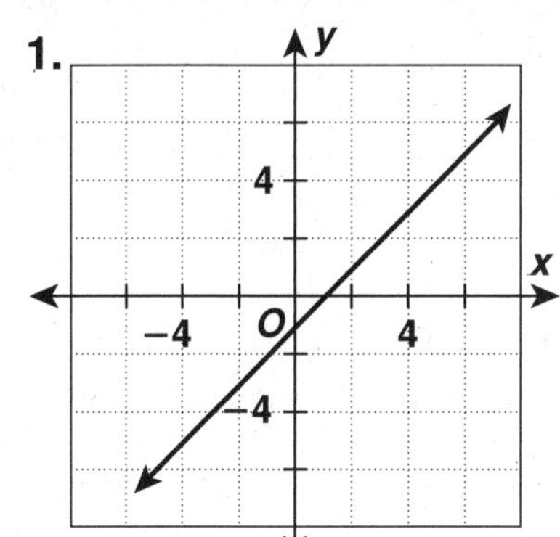

I linear Q nonlinear

2.

A linear Z nonlinear

3.

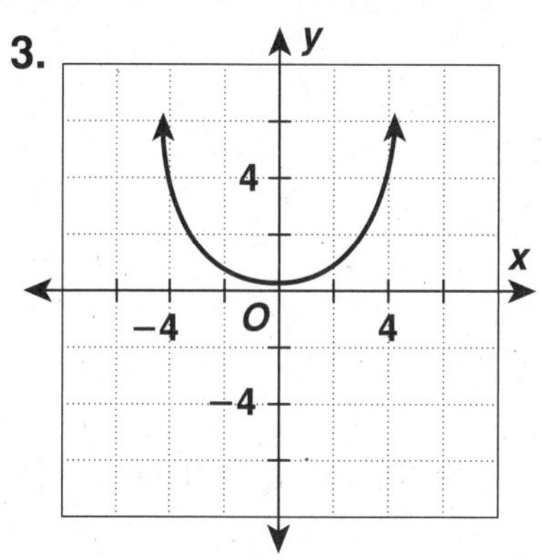

X linear N nonlinear

4.

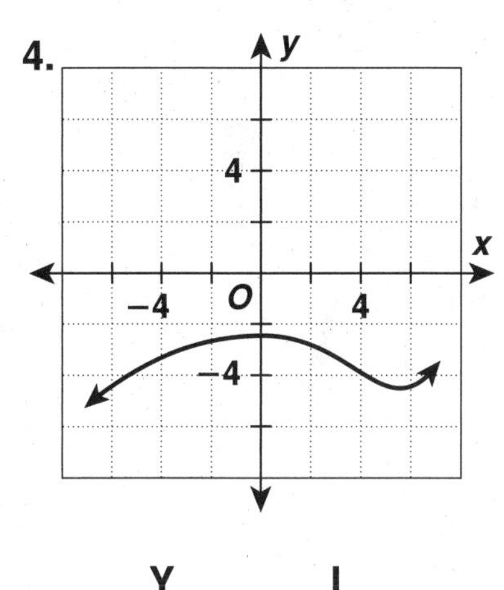

Y linear L nonlinear

5.

R linear V nonlinear

6.

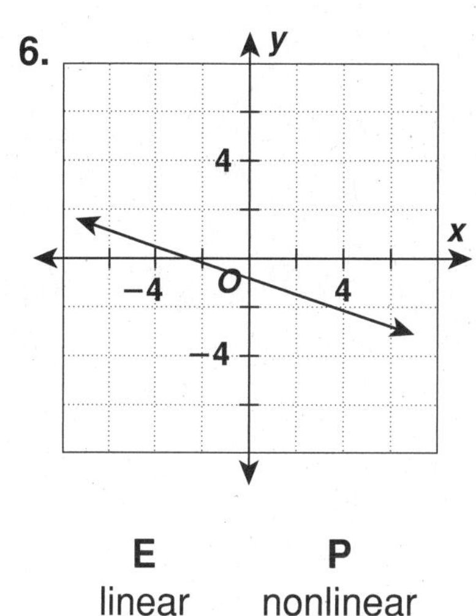

E linear P nonlinear

What do you call a dancing sheep?

A B ___ ___ – ___ ___ ___ ___ ___ ___ ___

2 2 4 4 6 5 1 3 2

Name ______________________ Date __________ Class __________

LESSON 12-2

Practice A

Slope of a Line

Find the slope of the line that passes through each pair of points.

1. (1, 0), (2, 4) ____________

2. (6, 2), (2, −2) ____________

3. (−1, 1), (4, 4) ____________

4. (−7, 4), (2, 1) ____________

5. (5, −3), (−2, −3) ____________

6. (−3, 2), (2, 7) ____________

Determine whether each graph shows a constant or variable rate of change. Explain your reasoning.

7.

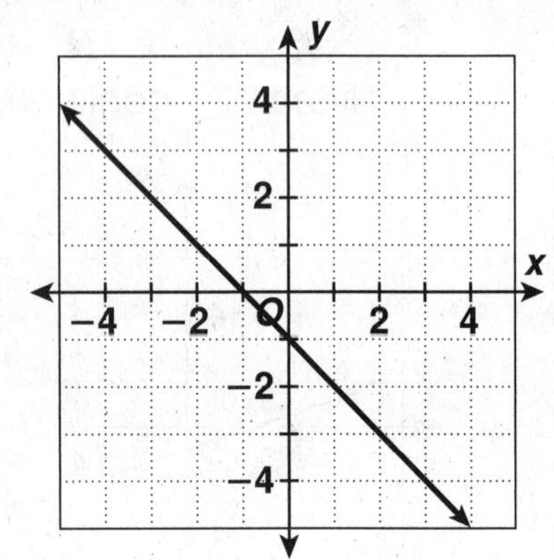

____

____

8.

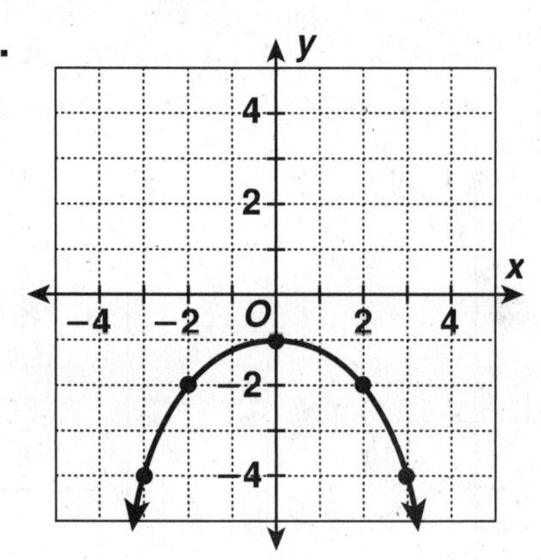

____

____

9.

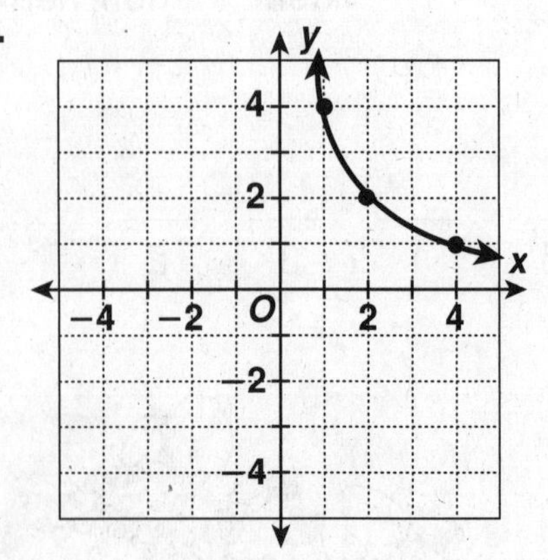

____

____

10.

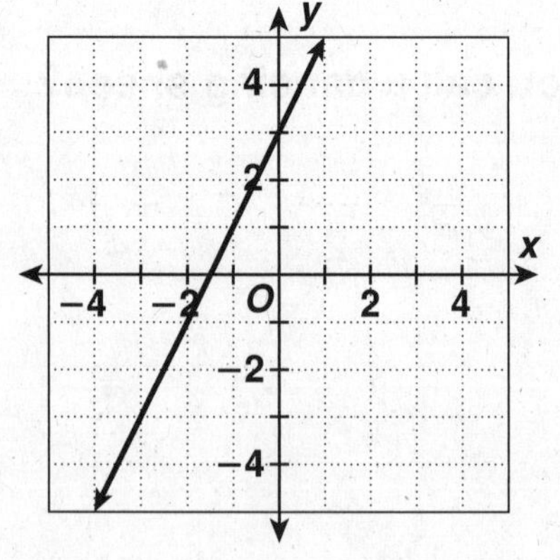

____

____

Name ______________________ Date __________ Class __________

LESSON **12-2**

Practice B
Slope of a Line

Find the slope of the line that passes through each pair of points.

1. (−2, −8), (1, 4) ____________
2. (−2, 0), (0, 4), ____________
3. (0, 4), (4, 4) ____________
4. (3, −6), (2, −4) ____________
5. (−3, 4), (3, −4) ____________
6. (3, 0), (0, −6), ____________
7. (3, 2), (3, −2) ____________
8. (−4, 4), (3, −1) ____________

Determine whether each graph shows a constant or variable rate of change. Explain your reasoning.

9.

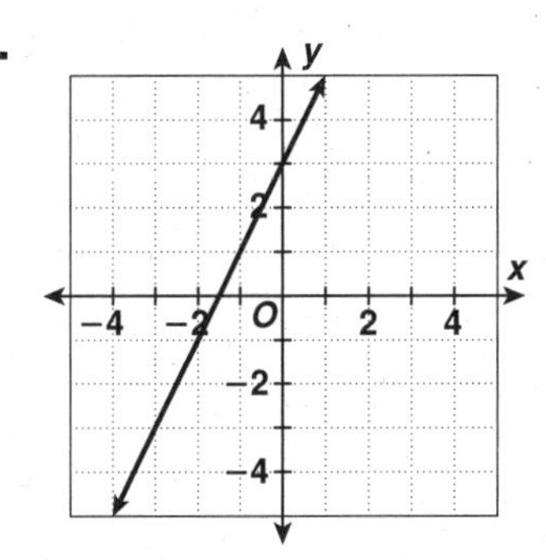

____

10.

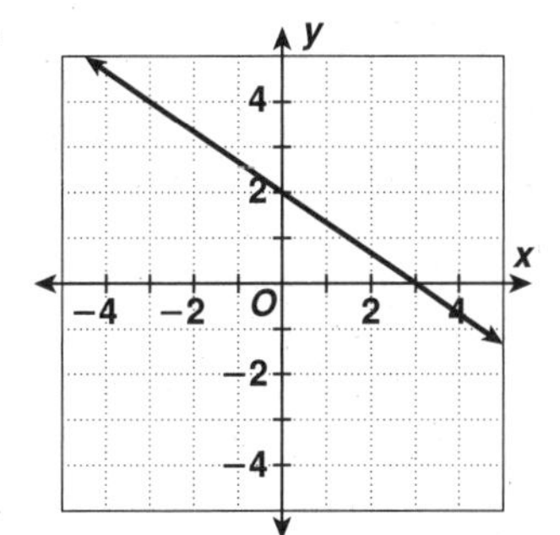

____

11.

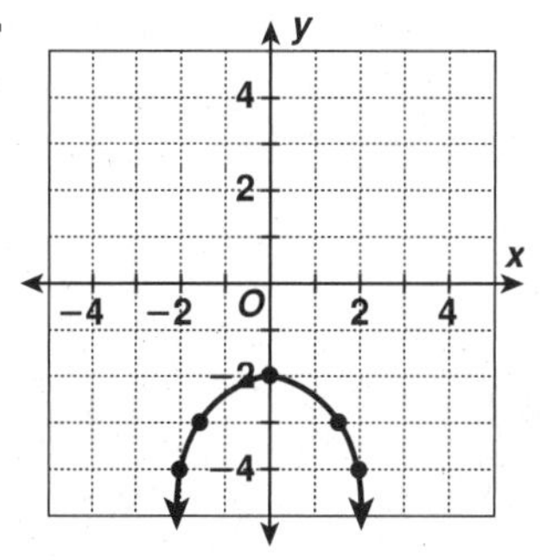

____

12. The table shows the distance Ms. Long had traveled as she went to the beach. Use the data to make a graph. Find the slope of the line and explain what it shows.

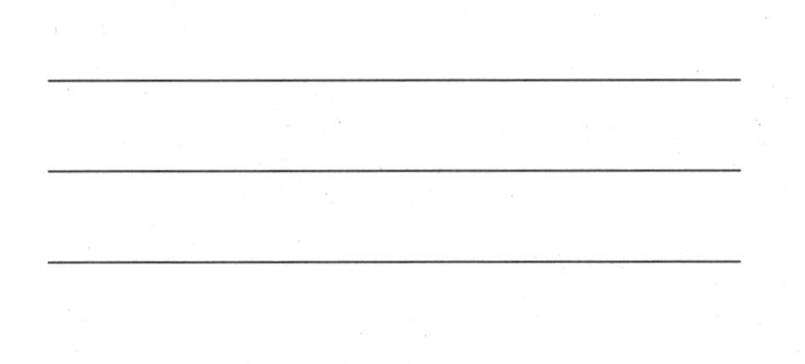

Time (min)	Distance (mi)
8	6
12	9
16	12
20	15

Distance Traveled (mi)

20
10
Miles
10
20
Time (min)

Name ______________________________ Date ____________ Class ____________

LESSON **12-2**

Practice C

Slope of a Line

For exercises 1–4, use the graph to find the slope of the line.

1.

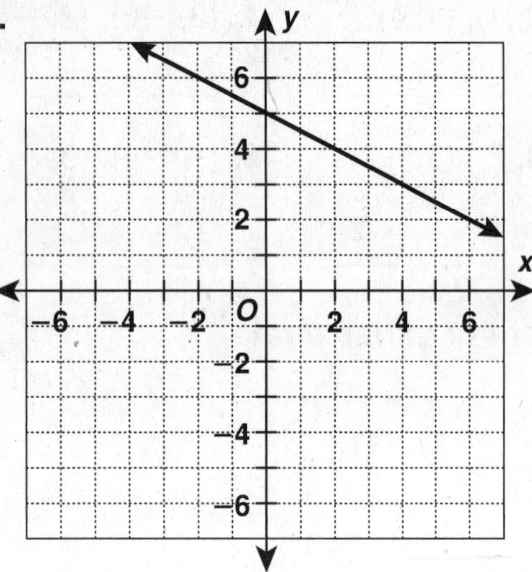

____

2.

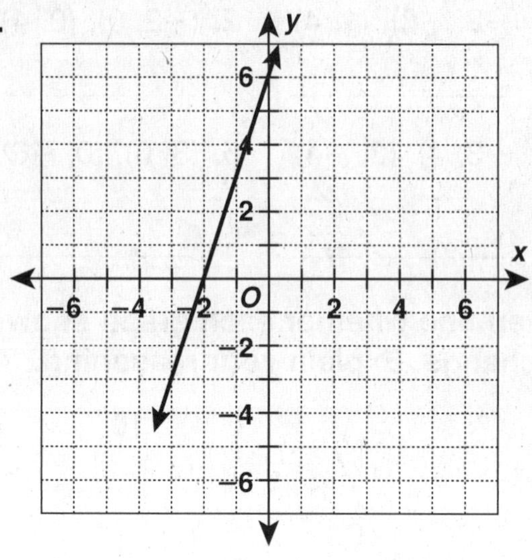

____

3.

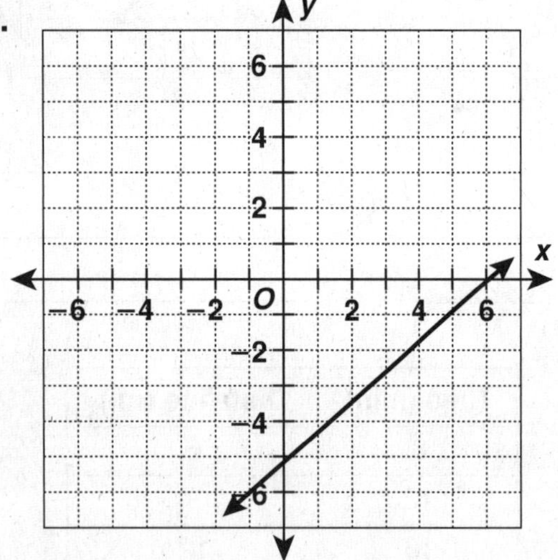

____

4.

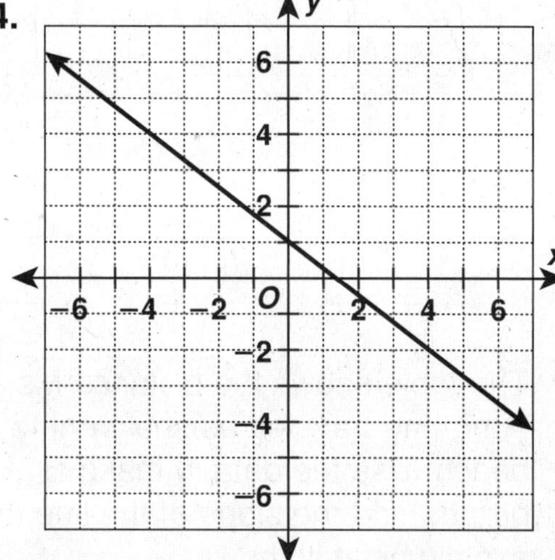

____

5. If a quadrilateral has vertices $A(-2, 2)$, $B(2, 3)$, $C(3, -4)$, and $D(-3, -2)$, find the slope of $\overline{AB}$, $\overline{BC}$, $\overline{CD}$, and $\overline{DA}$.

____

____

Name ______________________________ Date ___________ Class ___________

LESSON 12-2

Reteach

Slope of a Line

The **slope** of a line is a measure of its tilt, or slant.

The slope of a straight line is a constant ratio, the "rise over run," or the **vertical change** over the **horizontal change**.

You can find the slope of a line by comparing any two of its points. The vertical change is the difference between the two *y*-values. And the horizontal change is the difference between the two *x*-values.

$$\text{slope} = \frac{y_2 y_1}{x_2 x_1}$$

point *A*: (3, 2) point *B*: (4, 4)

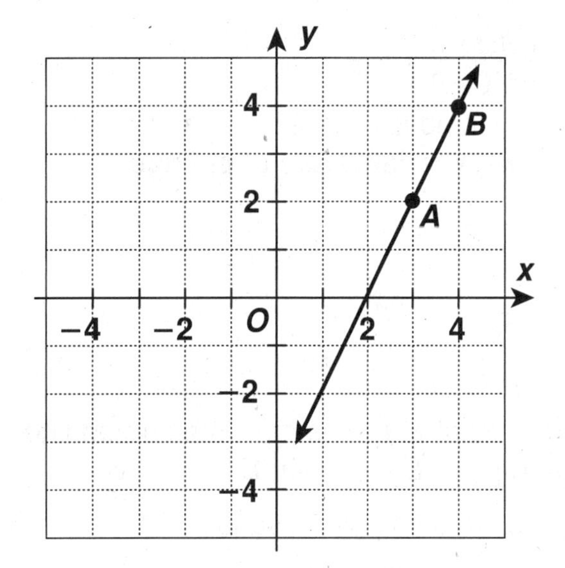

Make point *A* (x_1, y_1).
Make point *B* (x_2, y_2).

$$\text{slope} = \frac{4-2}{4-3}$$
$$= \frac{2}{1}, \text{ or } 2$$

So, the slope of the line is 2.

You can make point *A* (x_2, y_2) and point *B* (x_1, y_1).

$$\text{slope} = \frac{2-4}{3-4}$$
$$= \frac{-2}{-1}, \text{ or } 2$$

So, the slope remains 2.

Find the slope of the line that passes through each pair of points.

1. (1, 5) and (2, 6) ____________

2. (0, 3) and (2, 7) ____________

3. (2, 5) and (3, 4) ____________

4. (6, 9) and (2, 7) ____________

5. (6, 5) and (8, −1) ____________

6. (7, −4) and (4, −2) ____________

Name ______________________ Date ____________ Class ____________

LESSON 12-2

Reteach

Slope of a Line (continued)

A straight line has a constant slope, so it shows a **constant rate of change**. The same change in *y* always results in the same change in *x*.

From point *C* to point *B*:

$$\frac{\text{change in y}}{\text{change in } x} = \frac{2}{3}$$

From point *B* to point *A*:

$$\frac{\text{change in y}}{\text{change in } x} = \frac{2}{3}$$

A curved line doesn't have a constant slope, so it shows a **variable rate of change**.

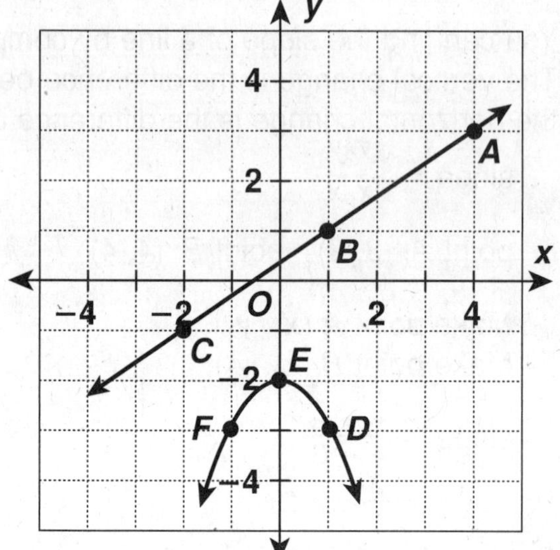

Between point *F* and point *E*, the curved line has a positive slope.
Between point *E* and point *D*, the curved line has a negative slope.

So, the curved line has a variable rate of change.

Determine whether each graph shows a constant or a variable rate of change. Write *constant* or *variable.*

7.

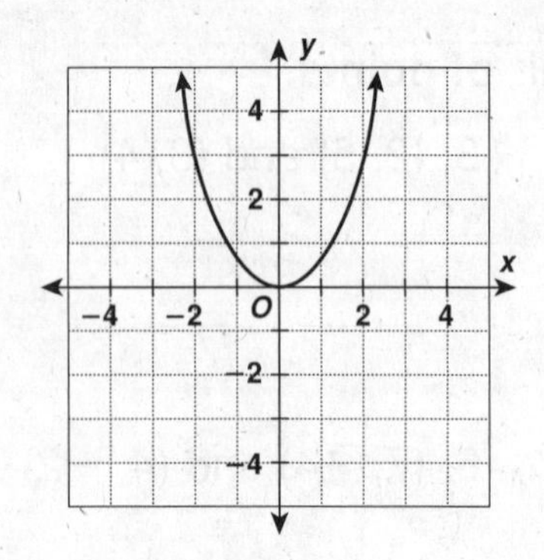

____

8.

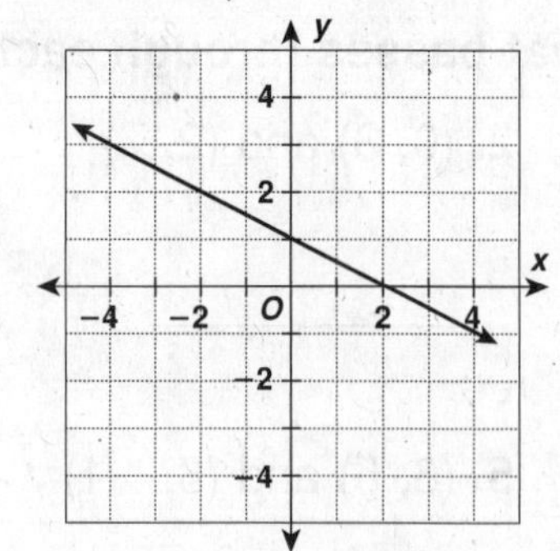

____

9.

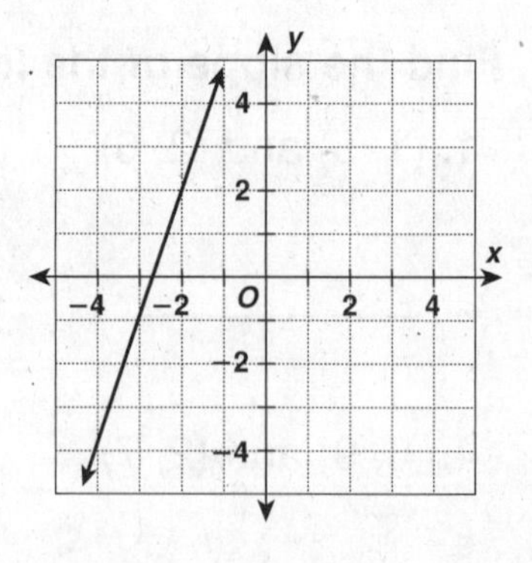

____

Name ______________________ Date __________ Class __________

LESSON 12-2

Challenge
Aligned?

1. Points A, B, and C are on the same line. Draw a conclusion about the slope between A and B and the slope between B and C.

____

2. Determine if the three points are collinear (lie on the same line).

a. $R(2, 5)$, $S(6, 15)$, $T(16, 18)$
slope between R and S =

____

slope between S and T =

____

R, S, T __________ collinear.

b. $J(0, -4)$, $K(1, -2)$, $L(3, 2)$
slope between J and K =

____

slope between K and L =

____

J, K, L __________ collinear.

3. Find the value of k so that $U(-5, -1)$, $V(-1, -5)$, and $W(5, k)$ are collinear.

a. Find the slope between U and V. ______________________________

b. Find the slope between V and W. ______________________________

c. Set the results of parts a and b equal to each other and solve for k. Justify your result.

______________________________ ______________________________

______________________________ ______________________________

______________________________ ______________________________

______________________________ ______________________________

4. The points $P(2, -3)$, $Q(2, 3)$ and $R(k, 0)$ are collinear. Find k. Justify your result.

____

____

Name ______________________ Date __________ Class __________

LESSON **12-2**

Problem Solving

Slope of a Line

Write the correct answer.

1. The state of Kansas has a fairly steady slope from the east to the west. At the eastern side, the elevation is 771 ft. At the western edge, 413 miles across the state, the elevation is 4039 ft. What is the approximate slope of Kansas?

____

2. The Feathered Serpent Pyramid in Teotihuacan, Mexico, has a square base. From the center of the base to the center of an edge of the pyramid is 32.5 m. The pyramid is 19.4 m high. What is the slope of each face of the pyramid?

____

3. On a highway, a 6% grade means a slope of 0.06. If a highway covers a horizontal distance of 0.5 miles and the elevation change is 184.8 feet, what is the grade of the road? (Hint: 5280 feet = 1 mile.)

____

4. The roof of a house rises vertically 3 feet for every 12 feet of horizontal distance. What is the slope, or pitch of the roof?

____

Use the graph for Exercises 5–8.

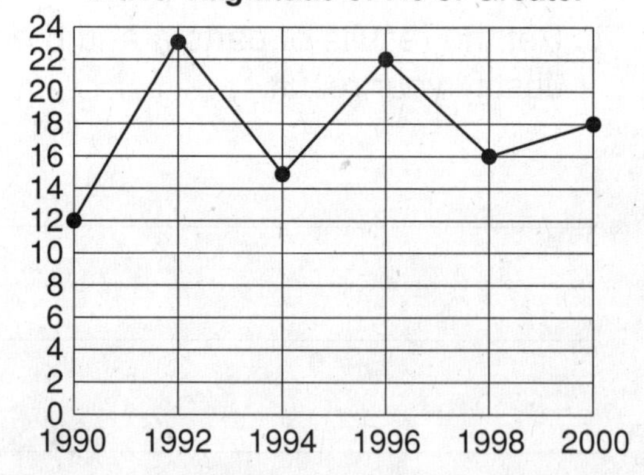

5. Find the slope of the line between 1990 and 1992.

A $\frac{2}{11}$ **C** $\frac{11}{2}$

B $\frac{35}{3982}$ **D** $\frac{11}{1992}$

6. Find the slope of the line between 1994 and 1996.

F $\frac{7}{2}$ **H** $\frac{2}{7}$

G $\frac{37}{3990}$ **J** $\frac{7}{1996}$

7. Find the slope of the line between 1998 and 2000.

A 1

B $\frac{1}{999}$

C $\frac{1}{1000}$

D 2

8. What does it mean when the slope is negative?

F The number of earthquakes stayed the same.

G The number of earthquakes increased.

H The number of earthquakes decreased.

J It means nothing.

Name ______________________ Date __________ Class __________

LESSON 12-2

Reading Strategies

Use a Graphic Organizer

Slope of a Line

Definition **Slope** is a measure of the slant of a line.	Slope is a ratio. (vertical change compared to horizontal change; $\frac{\text{vertical}}{\text{horizontal}}$)
Lines with Nonzero Slope Positive slope: The line slants upward from left to right. 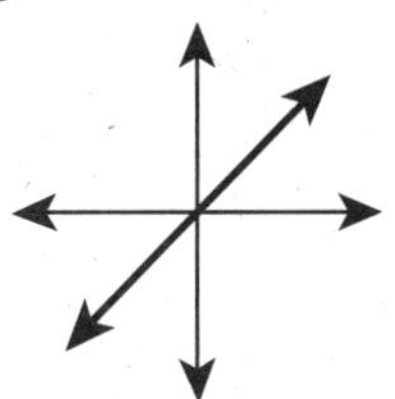 Negative slope: The line slants downward from left to right. 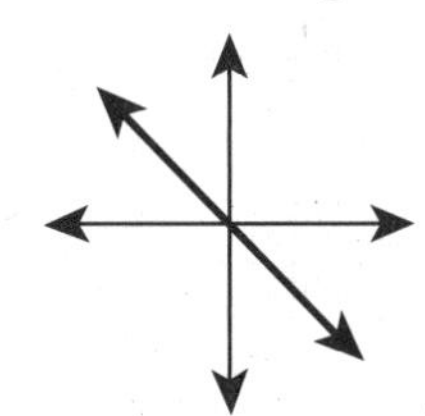	**Horizontal and Vertical Lines** Zero slope: Horizontal lines have a slope of 0. 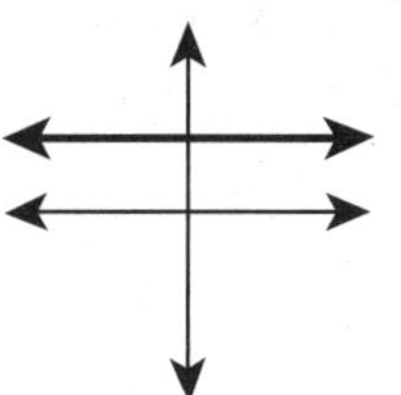 Undefined slope: Vertical lines have an undefined slope. 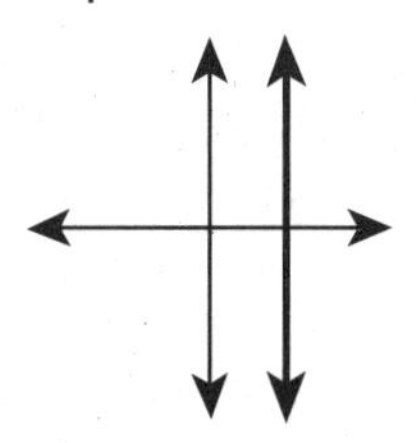

Use the graphic organizer to answer the following questions.

1. What do you call the slant of a line?

____

2. Write the ratio that is used to describe slope.

____

3. How can you tell if a line has positive slope?

____

4. How you can tell if a line has negative slope?

____

5. What kind of line has a slope of 0?

____

Name ______________________ Date __________ Class __________

LESSON **12-2**

Puzzles, Twisters & Teasers

A Slippery Slope!

Determine which kind of slope each line has. Then use the letters of the correct answers to solve the riddle.

1.

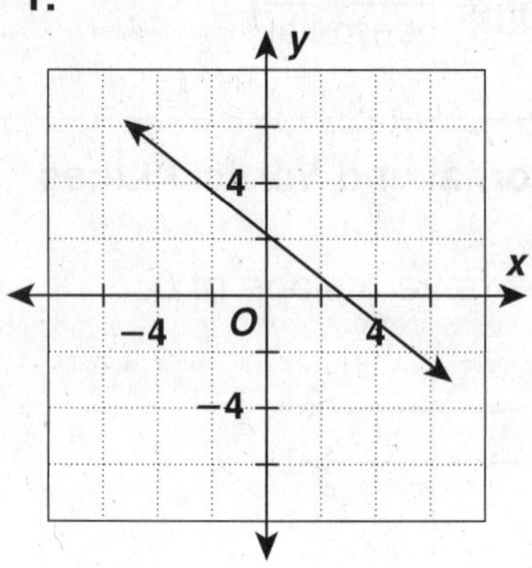

Positive **B** Negative **O**
Zero **D** Undefined **L**

2.

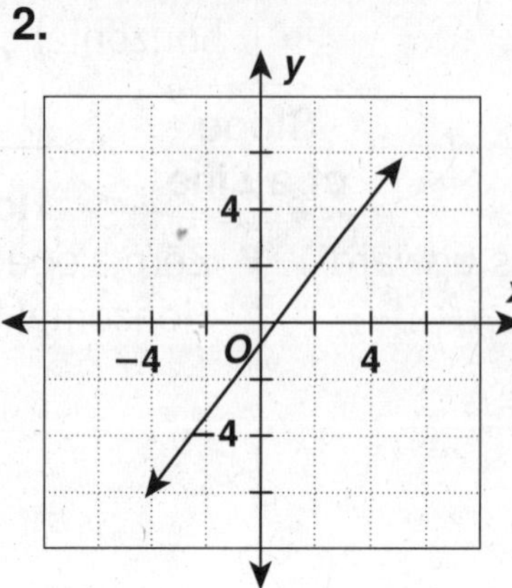

Positive **R** Negative **C**
Zero **Q** Undefined **D**

3.

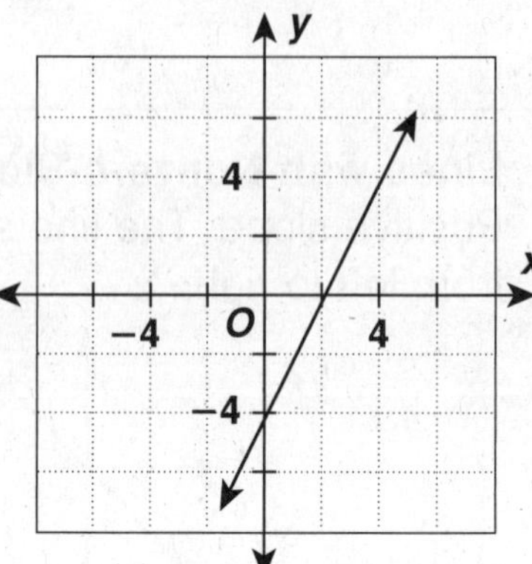

Positive **T** Negative **N**
Zero **G** Undefined **N**

4.

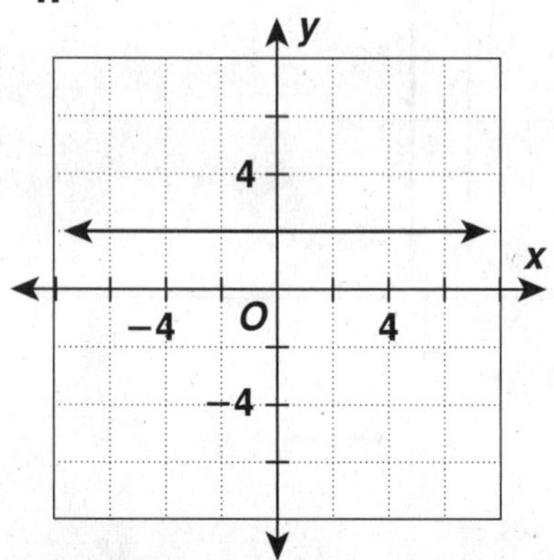

Positive **A** Negative **T**
Zero **H** Undefined **V**

5.

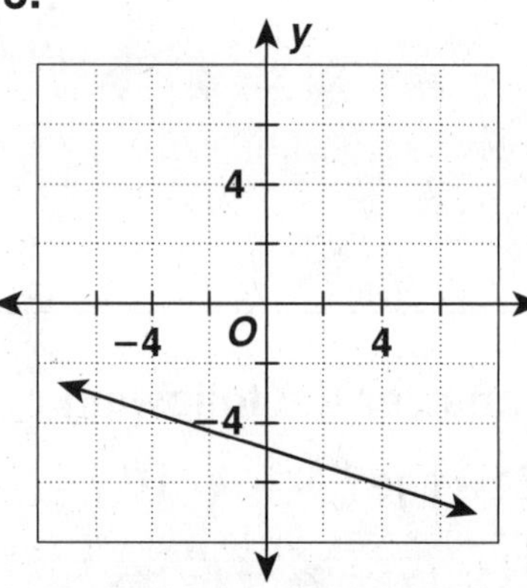

Positive **P** Negative **U**
Zero **F** Undefined **Q**

6.

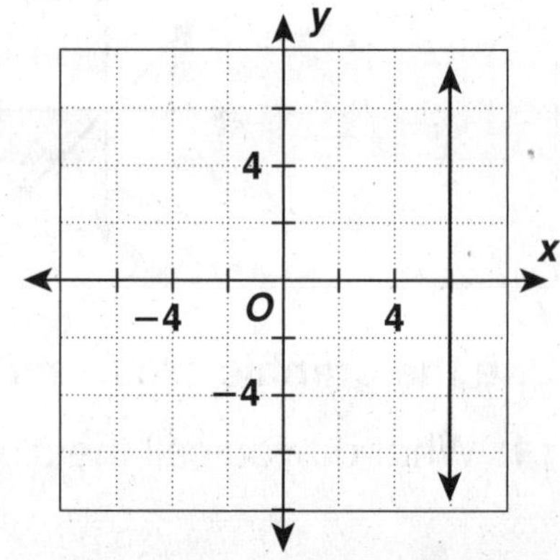

Positive **J** Negative **W**
Zero **X** Undefined **Y**

What time do you go to the dentist?

____ ____ ____ ____ ____ – ____ ____ ____ ____ ____
3 1 1 3 4   4 5 2 3 6

Name ______________________ Date __________ Class __________

LESSON **11-3**

Practice A

Using Slopes and Intercepts

1. Name the ordered pair if the x-intercept is -2. ____________

2. Name the ordered pair if the y-intercept is 8. ____________

3. In the ordered pair (9, 0), what is the x-intercept? ____________

4. In the ordered pair (0, 0), what is the relationship of the x-intercept and y-intercept? ____________

Find the x-intercept and y-intercept of each line. Use the intercepts to graph the equation.

5. $x + y = 5$

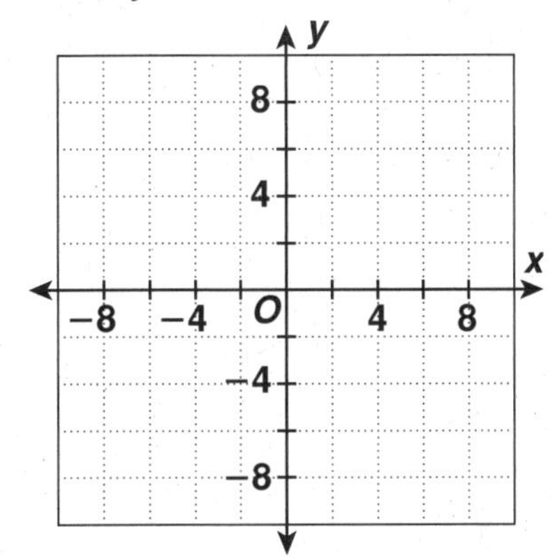

____

____

6. $2x - y = 6$

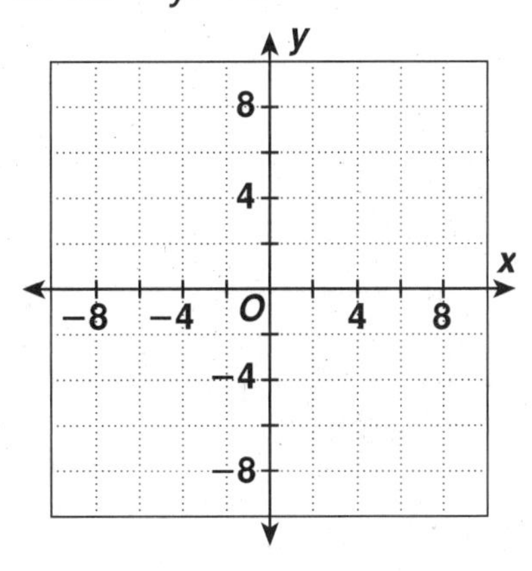

____

____

Write each equation in slope-intercept form, and then find the slope and y-intercept.

7. $2x + y = -5$

____

____

8. $x - y = 10$

____

____

9. $x - 2y = 4$

____

____

Write the equation of the line that passes through each pair of points in slope-intercept form.

10. (1, 2), (−1, 0)

____

11. (1, −3), (−1, 1)

____

12. (1, 1), (−3, −3)

____

Name ______________________ Date ____________ Class ____________

LESSON 12-3

Practice B

Using Slopes and Intercepts

Find the x-intercept and y-intercept of each line. Use the intercepts to graph the equation.

1. $x - y = -3$

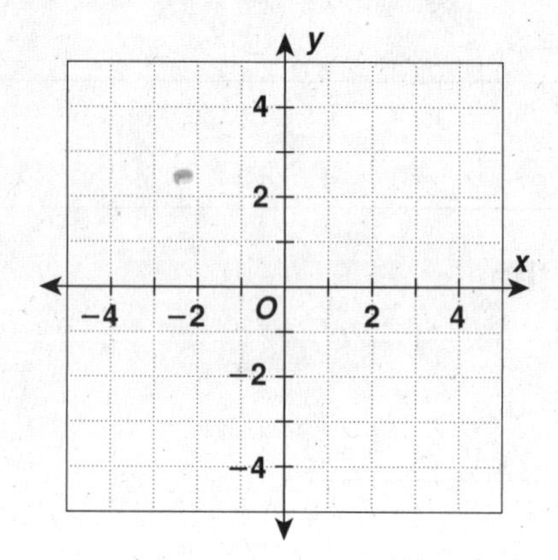

2. $2x + 3y = 12$

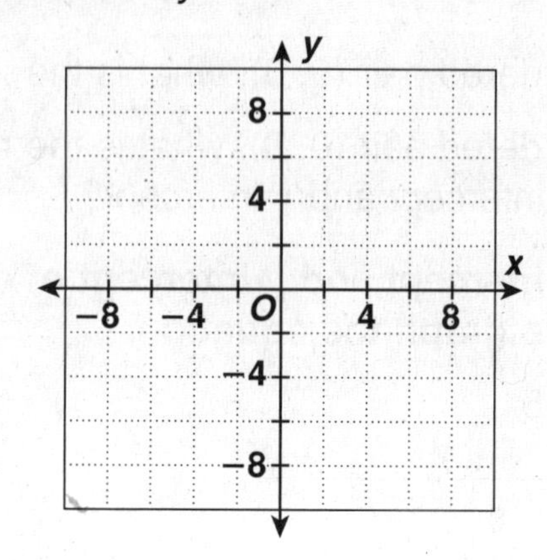

____

____

Write each equation in slope-intercept form, and then find the slope and y-intercept.

3. $3x + y = 0$

4. $2x - y = -15$

5. $x - 5y = 10$

Write the equation of the line that passes through each pair of points in slope-intercept form.

6. (3, 4), (4, 6)

7. (−1, −1), (2, −10)

8. (6, 5), (−9, −20)

9. A pizzeria charges \$8 for a large cheese pizza, plus \$2 for each topping. The total cost for a large pizza is given by the equation $C = 2t + 8$, where t is the number of toppings. Identify the slope and y-intercept, and use them to graph the equation for t between 0 and 5 toppings.

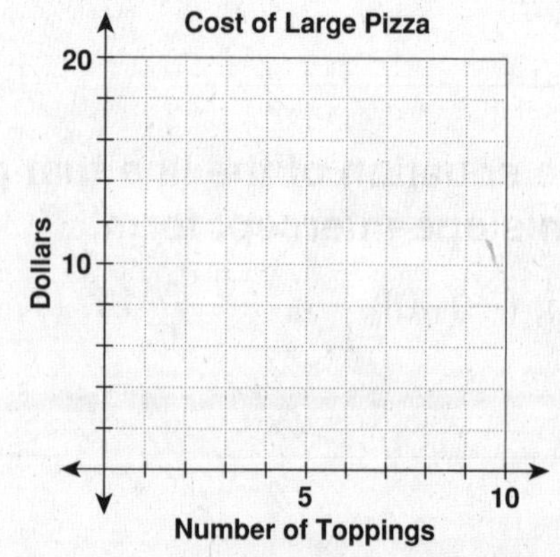

____

Name ______________________ Date __________ Class __________

LESSON 12-3

Practice C

Using Slopes and Intercepts

Find the *x*-intercept and *y*-intercept of each line. Use the intercepts to graph the equation.

1. $3x - 2y = 6$

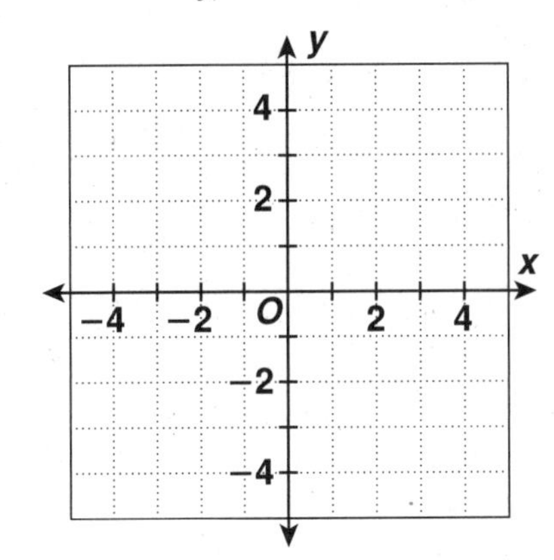

2. $5x + 4y = 20$

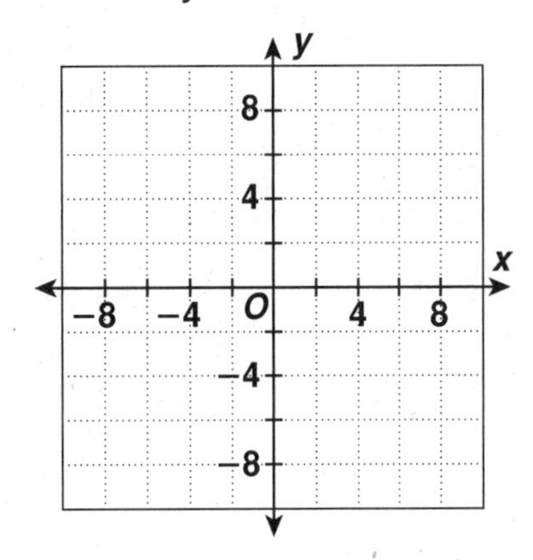

______________________ ______________________

Write each equation in slope-intercept form, and then find the slope and y-intercept.

3. $y - 3x = -10$

4. $3y - 2x = 9$

5. $6y - 2x = -\frac{1}{2}$

______________ ______________ ______________

Write the equation of the line that passes through each pair of points in slope-intercept form.

6. (3, 4), (−1, −4)

7. (6, 10), (12, 14)

8. (9, −3), (9, 5)

______________ ______________ ______________

9. A home improvement warehouse charges a $60 delivery fee. A customer wants to purchase a number of pieces of lumber that cost $5 a piece. Write an equation in slope-intercept form, where *C* is the total cost of the delivered lumber and *x* represents the number of pieces of lumber purchased. Graph the equation for *x* between 1 and 5 pieces.

____

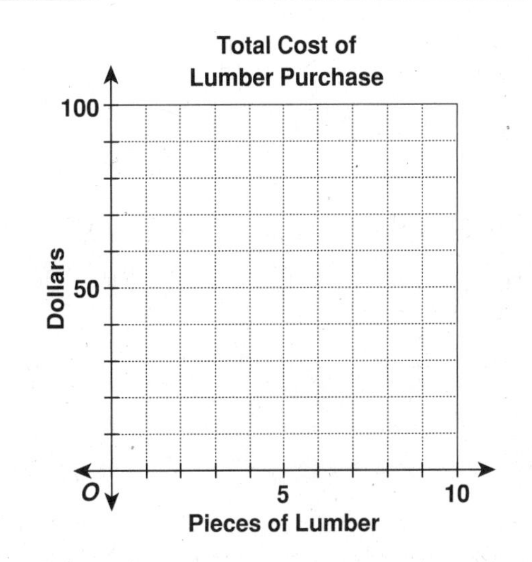

Name ______________________ Date __________ Class __________

LESSON 12-3

Reteach

Using Slopes and Intercepts

x-intercept: the x-coordinate of the point at which a line crosses the x-axis

y-intercept: the y-coordinate of the point at which a line crosses the y-axis

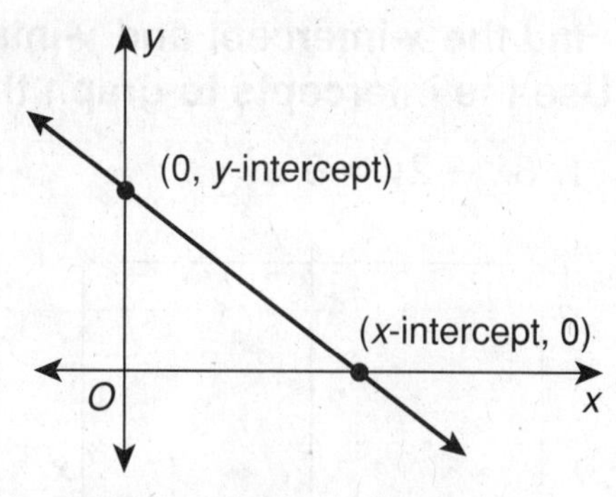

You can find the intercepts of a line from its equation. Then you can use the intercepts to graph the line.

Find the intercepts of the line $3x + 4y = 24$.

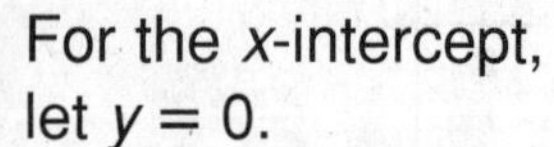

For the x-intercept, let $y = 0$.

$$3x + 4y = 24$$
$$3x + 4(0) = 24$$
$$3x + 0 = 24$$
$$3x = 24$$
$$\frac{3x}{3} = \frac{24}{3}$$
$$x = 8$$

The x-intercept is 8.

For the y-intercept, let $x = 0$.

$$3x + 4y = 24$$
$$3(0) + 4y = 24$$
$$0 + 4y = 24$$
$$4y = 24$$
$$\frac{4y}{4} = \frac{24}{4}$$
$$y = 6$$

The y-intercept is 6.

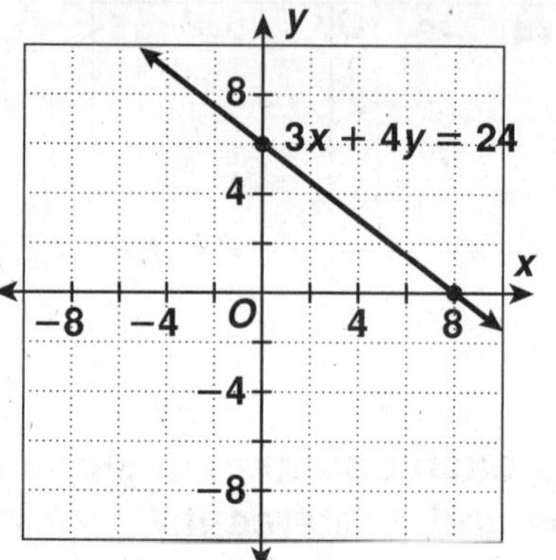

Find the intercepts of each line.
Draw both graphs on the same grid.

1. $2x + 3y = 12$

for x-intercept

$2x + 3(\_\_\_) = 12$

____

____

$x =$ ____

for y-intercept

$2(\_\_\_) + 3y = 12$

____

____

$y =$ ____

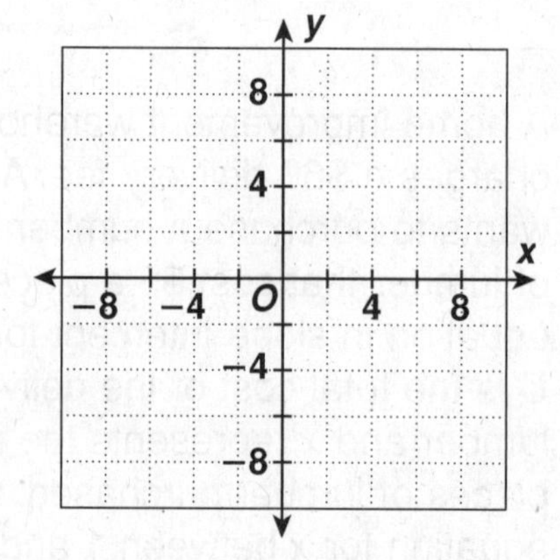

2. $6y - 3x = 6$

for x-intercept

$6(\_\_\_) - 3x = 6$

____

____

$x =$ ____

for y-intercept

$6y - 3(\_\_\_) = 6$

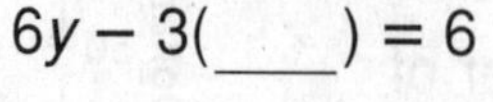

____

____

$y =$ ____

Name ______________________ Date __________ Class __________

LESSON 12-3

Reteach

Using Slopes and Intercepts (continued)

slope-intercept form

$y = mx + b$

slope (m) y-intercept (b)

In this form, the coefficient of x is the slope and the constant term is the y-intercept.

To rewrite an equation in slope-intercept form, *isolate y.*

$$2x + 3y = -12$$
$$-2x \qquad -2x \qquad \text{Subtract } 2x.$$
$$3y = -2x - 12$$
$$\frac{3y}{3} = \frac{-2}{3}x - \frac{12}{3} \qquad \text{Divide by 3.}$$
$$y = \frac{-2}{3}x - 4$$

So, $m = -\frac{2}{3}$ and $b = -4$.

Write each equation in slope-intercept form and then find the slope and y-intercept.

3. $3y = 4x + 15$

$\frac{3y}{} = \frac{4}{}x + \frac{15}{}$ Divide by 3.

____

____

4. $3x - 2y = 6$

______ ______ Subtract $3x$.

____

____

Given two points of a line, you can write its equation.

(2, 5) and (−1, −4)

$$\text{slope} = \frac{y_2 y_1}{x_2 x_1} = \frac{-4-5}{-1-2} = \frac{-9}{-3} = 3$$

To find b, substitute the slope and the values from one of the points into the slope-intercept equation.

$$y = mx + b \rightarrow 5 = 3(2) + b$$
$$5 = 6 + b$$
$$-1 = b$$

So, the equation for the line that passes through (2, 5) and (−1, −4) is $y = 3x - 1$.

Write the equation of the line that passes through each pair of points in slope-intercept form.

5. (2, 11) and (0, 3)

____

6. (−1, 3) and (4, −2)

____

7. (10, 1) and (6, −1)

____

Name ______________________ Date __________ Class __________

LESSON 12-3

Challenge

Another View

The **intercepts** of a line are the points where the line crosses the coordinate axes.

When an equation was in standard form, you found the intercepts by setting one variable and then the other equal to zero.

1. Find the intercepts of the line whose equation is $3x + 5y = 15$.

For the x-intercept, let $y = 0$.	For the y-intercept, let $x = 0$.

The x-intercept is ______. The y-intercept is ______.

When an equation is in standard form, you can also find the intercepts by dividing both sides by the constant (on the right side).

2. Divide both sides of the equation $3x + 5y = 15$ by 15, and simplify. Compare the results to those obtained in Question 1.

____

____

3. Using b to represent the y-intercept and a to represent the x-intercept, write an equation that generalizes the observation you made in Question 2.

____

4. a. Using the form of the equation you wrote in Question 3, find the intercepts of the linear equation $2x + 3y = 24$.

____

The x-intercept is ______ and the y-intercept is ______.

b. Check your result by using the first method to find the intercepts.

______________ ______________

______________ ______________

______________ ______________

Name ______________________ Date ____________ Class ____________

LESSON 12-3

Problem Solving

Using Slopes and Intercepts

Write the correct answer.

1. Jaime purchased a $20 bus pass. Each time she rides the bus, $1.25 is deducted from the pass. The linear equation $y = -1.25x + 20$ represents the amount of money on the bus pass after x rides. Identify the slope and the x- and y-intercepts. Graph the equation at the right.

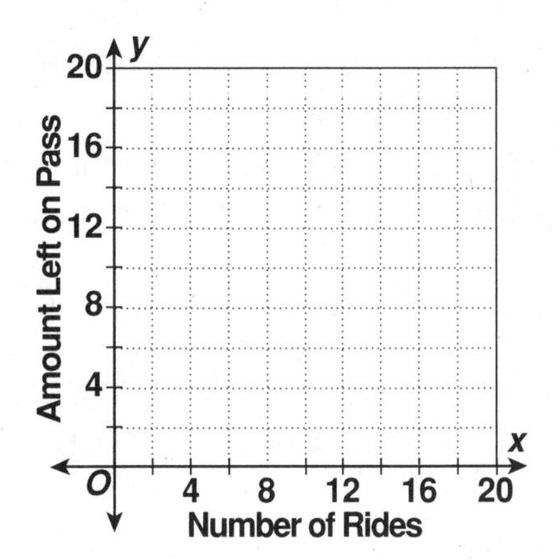

____

____

2. The rent charged for space in an office building is related to the size of the space rented. The rent for 600 square feet of floor space is $750, while the rent for 900 square feet is $1150. Write an equation for the rent y based on the square footage of the floor space x.

____

Choose the letter of the correct answer.

3. A limousine charges $35 plus $2 per mile. Which equation shows the total cost of a ride in the limousine?

   A $y = 35x + 2$ **C** $y = 2x - 35$

   B $y = 2x + 35$ **D** $2x + 35y = 2$

4. A newspaper pays its delivery people $75 each day plus $0.10 per paper delivered. Which equation shows the daily earnings of a delivery person?

   F $y = 0.1x + 75$ **H** $x + 0.1y = 75$

   G $y = 75x + 0.1$ **J** $0.1x + y = 75$

5. A friend gave Ms. Morris a $50 gift card for a local car wash. If each car wash costs $6, which equation shows the number of dollars left on the card?

   A $50x + 6y = 1$ **C** $y = -6x + 50$

   B $y = 6x + 50$ **D** $y = 6x - 50$

6. Antonio's weekly allowance is given by the equation $A = 0.5c + 10$, where c is the number of chores he does. If he received $16 in allowance one week, how many chores did he do?

   F 10 **H** 14

   G 12 **J** 15

Name ______________________ Date ___________ Class ___________

LESSON 12-3

Reading Strategies

Use a Visual Model

Refer to the coordinate plane at the right. Find the point where the line crosses the *x*-axis. This point is called the ***x*-intercept.**

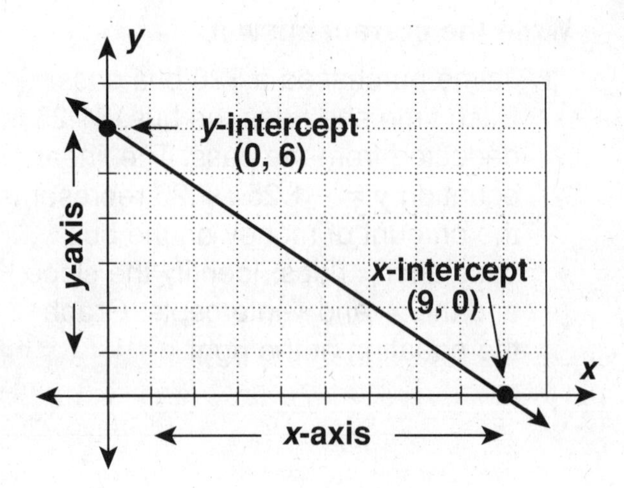

1. What is the *y*-value of the ordered pair for this point?

____

Find the point where the line crosses the *y*-axis. This point is called the ***y*-intercept.**

2. What is the *x*-value of the ordered pair for this point?

____

3. Which axis does the line cross at the *x*-intercept?

____

4. Name the ordered pair for the point where the line crosses the *x*-axis.

____

5. Which axis does the line cross at the *y*-intercept?

____

6. Name the ordered pair for the point where the line crosses the *y*-axis.

____

Name ______________________ Date __________ Class __________

LESSON 12-3

Puzzles, Twisters & Teasers

Word Bath!

Circle words from the list in the word search (horizontally, vertically or diagonally). Find a word that answers the riddle.

intercept	slope	form	graph	coordinate
axis	point	line	rate	change

```
C H A N G E L I N E F Q
O L H J U I R N R T O W
O P E E R T Y T Y U R E
R F G A S K I E O M M P
D C V B N A L R P L M O
I W E R T Y U C N K O I
N D F G H J K E B J I N
A X I S S L O P E U H T
T A S D F R A T E V G Y
E G R A P H C X Z A S D
```

Why did the robber take a bath?

Because he wanted to make a ____________________ getaway.

Name ______________________ Date ____________ Class ____________

LESSON 12-4

Practice A

Point-Slope Form

Use the point-slope form of each equation to identify the slope of the line.

1. $y - 1 = 2(x - 3)$
2. $y + 4 = -1(x + 7)$
3. $y - 7 = -3(x + 8)$

______________ ______________ ______________

Use the point-slope form of each equation to identify a point each line passes through.

4. $y - 1 = -2(x - 4)$
5. $y + 3 = -5(x - 1)$
6. $y + 5 = -2(x + 6)$

______________ ______________ ______________

Use the point-slope form of each equation to identify a point the line passes through and the slope of the line.

7. $y - 9 = 2(x - 3)$
8. $y + 6 = -3(x - 1)$
9. $y + 1 = -7(x + 2)$

______________ ______________ ______________

______________ ______________ ______________

10. $y + 2 = -6(x - 7)$
11. $y + 6 = -5(x + 9)$
12. $y - 3 = \frac{1}{3}(x + 9)$

______________ ______________ ______________

______________ ______________ ______________

Write the point-slope form of the equation with the given slope that passes through the indicated point.

13. the line with slope -4 passing though (5, 4)

____

14. the line with slope 2 passing through (-1, -2)

____

Name ________________________ Date ____________ Class ____________

LESSON 12-4

Practice B

Point-Slope Form

Use the point-slope form of each equation to identify a point the line passes through and the slope of the line.

1. $y - 2 = 4(x - 1)$

____

____

2. $y + 1 = 2(x - 3)$

____

____

3. $y - 4 = -3(x + 1)$

____

____

4. $y + 5 = -2(x + 6)$

____

____

5. $y + 4 = -9(x + 3)$

____

____

6. $y - 7 = -7(x - 7)$

____

____

7. $y - 10 = 6(x - 8)$

____

____

8. $y + 12 = 2.5(x + 4)$

____

____

9. $y + 8 = \frac{1}{2}(x - 3)$

____

____

Write the point-slope form of the equation with the given slope that passes through the indicated point.

10. the line with slope −1 passing through (2, 5)

____

11. the line with slope 2 passing through (−1, 4)

____

12. the line with slope 4 passing through (−3, −2)

____

13. the line with slope 3 passing through (7, −6)

____

14. the line with slope −3 passing through (−6, 4)

____

15. the line with slope −2 passing through (5, 1)

____

16. Michael was driving at a constant speed of 60 mph when he crossed the Sandy River. After 1 hour, he passed a highway marker for mile 84. Write an equation in point-slope form, and find which highway marker he will pass 90 minutes after crossing the Sandy River.

____

Name ______________________ Date __________ Class __________

LESSON 12-4

Practice C

Point-Slope Form

Write the point-slope form of the equation with the given slope that passes through the indicated point.

1. the line with slope $\frac{1}{2}$ passing through $(-4, 8)$

____

2. the line with slope 7 passing through $(\frac{1}{3}, -6)$

____

3. the line with slope 2.6 passing through $(7.8, 4.5)$

____

4. the line with slope $\frac{5}{3}$ passing through $(2, 5)$

____

5. the line with slope $-\frac{3}{4}$ passing through $(\frac{1}{4}, \frac{1}{5})$

____

6. the line with slope -9 passing through $(-\frac{2}{3}, -9)$

____

The slopes of parallel lines are equal. The slopes of perpendicular lines are negative reciprocals. (If line *A* has a slope of 2 and line *A* is perpendicular to line *B*, then the slope of line *B* is $-\frac{1}{2}$.)

Write the point-slope form of each line described below.

7. the line parallel to $y = 5x - 1$ that passes through $(-2, 7)$

____

8. the line perpendicular to $y = 3x + 6$ that passes through $(-1, 0)$

____

9. the line perpendicular to $y = -\frac{2}{3}x$ that passes through $(-5, -5)$

____

10. the line parallel to $y = \frac{3}{4}x + 8$ that passes through $(-1, -9)$

____

11. A school librarian is packing up books for the summer. The boxes will hold either 6 English books and 18 math books, or 11 English books and 14 math books. Let *x* equal the number of English books and *y* equal the number of math books. Write two different equations in point-slope form using this information.

____

Name ______________________ Date __________ Class __________

LESSON 12-4

Reteach

Point-Slope Form

$$y - y_1 = m(x - x_1)$$

slope (arrow to m)

(x_1, y_1) are the coordinates of a known point on the line.

If a minus sign precedes a coordinate value, the coordinate is positive.

$y - 3 = 7(x - 1)$

(1, 3) is on the line; slope $m = 7$

If a plus sign precedes a coordinate value, the coordinate is negative.

$y + 3 = 7(x + 1)$

(−1, −3) is on the line; slope $m = 7$

Identify the slope of each line and a point it passes through.

1. $y + 2 = 5(x - 3)$

$m =$ ________

Which sign for each coordinate? ________

Coordinates of a point on the line: ________

2. $y - 4 = -3(x + 5)$

$m =$ ________

Which sign for each coordinate? ________

Coordinates of a point on the line: ________

To write an equation for the line with slope −4 that passes through (6, −2), substitute $m = -4$, $x_1 = 6$, $y_1 = -2$ into the point-slope form.

$$y - y_1 = m(x - x_1)$$
$$y - (-2) = -4(x - 6)$$
$$y + 2 = -4(x - 6)$$

Write the point-slope form of the equation with the given slope that passes through the given point.

3. $m = 3$; $(x_1, y_1) = (7, 2)$

$y - y_1 = m(x - x_1)$

$y -$ ______ $=$ ______ $(x -$ ______ $)$

4. $m = -5$; $(x_1, y_1) = (2, 6)$

$y - y_1 = m(x - x_1)$

$y -$ ______ $=$ ______ $(x -$ ______ $)$

5. $m = \frac{1}{2}$; $(x_1, y_1) = (-8, 1)$

$y - y_1 = m(x - x_1)$

$y -$ ______ $=$ ______ $(x$ ______ $)$

6. $m = -\frac{3}{4}$; $(x_1, y_1) = (0, -1)$

$y - y_1 = m(x - x_1)$

y ______ $=$ ______ $(x -$ ______ $)$

Name ______________________ Date ____________ Class ____________

LESSON 12-4

Challenge

So Everyone Gets the Same Answer

The **standard form** of a line is $Ax + By = C$ where A, B, and C are *real numbers*.

To write an equation of a line, you need to know two pieces of information.

When the slope and the y-intercept are known, use $y = mx + b$.
When the slope and a point on the line are known, use $y - y_1 = m(x - x_1)$.

You can use either the slope-intercept form or the point-slope form to write an equation in standard form.

Write an equation in standard form for the line that contains side $\overline{AB}$ of triangle ABC.

Use $A(1, 0)$ and $B(4, 5)$ to find the slope of $\overleftrightarrow{AB}$. $\quad m = \frac{5 - 0}{4 - 1} = \frac{5}{3}$

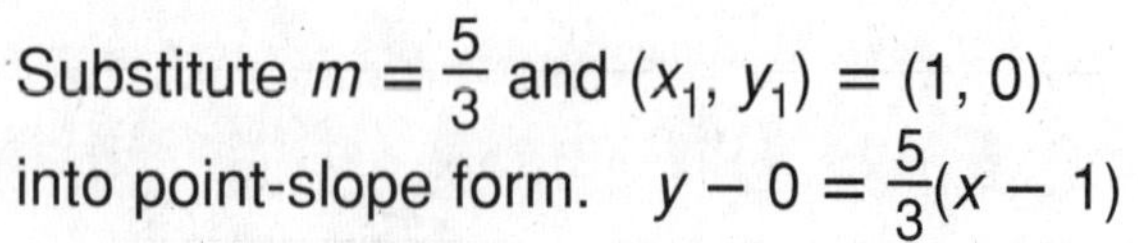

Substitute $m = \frac{5}{3}$ and $(x_1, y_1) = (1, 0)$ into point-slope form. $\quad y - 0 = \frac{5}{3}(x - 1)$

Write the equation in standard form.

clear fractions	$3y = 5(x - 1)$
distribute	$3y = 5x - 5$
add and subtract	$5x - 3y = 5$

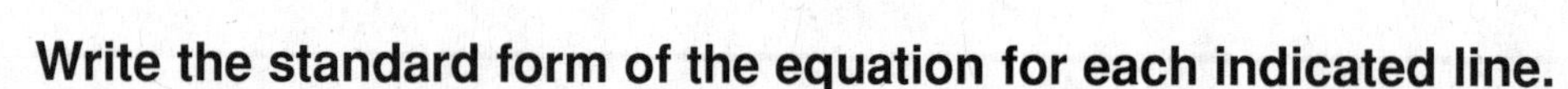

Write the standard form of the equation for each indicated line.

1.

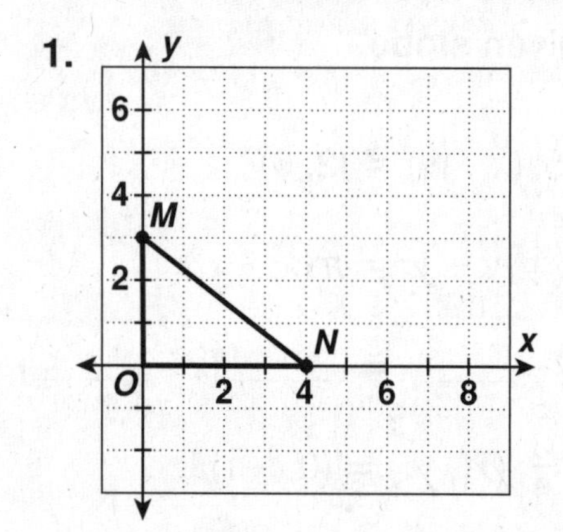

$\overleftrightarrow{MN}$ of right triangle MNO

____

2.

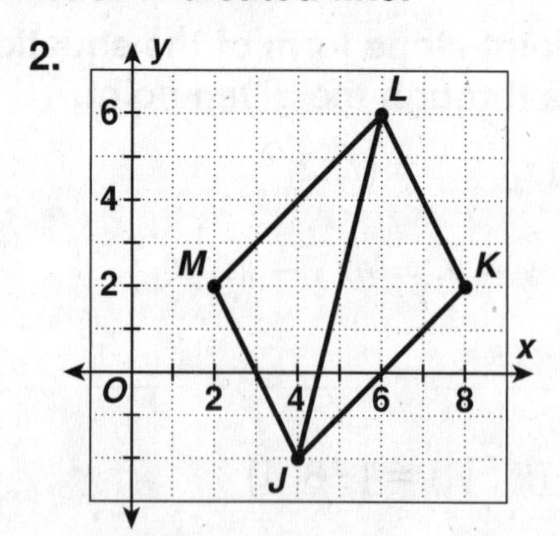

$\overleftrightarrow{JL}$ of parallelogram $JKLM$

____

Name ______________________ Date __________ Class __________

LESSON 12-4

Problem Solving

Point-Slope Form

Write the correct answer.

1. A 1600 square foot home in Houston will sell for about \$102,000. The price increases about \$43.41 per square foot. Write an equation that describes the price y of a house in Houston, based on the square footage x.

____

2. Write the equation in Exercise 1 in slope-intercept form.

____

3. Wind chill is a measure of what temperature feels like with the wind. With a 25 mph wind, 40°F will feel like 29°F. Write an equation in point-slope form that describes the wind chill y based on the temperature x, if the slope of the line is 1.337.

____

4. With a 25 mph wind, what does a temperature of 0°F feel like?

____

From 2 to 13 years, the growth rate for children is generally linear. Choose the letter of the correct answer.

5. The average height of a 2-year old boy is 36 inches, and the average growth rate per year is 2.2 inches. Write an equation in point-slope form that describes the height of a boy y based on his age x.

   A $y - 36 = 2(x - 2.2)$

   B $y - 2 = 2.2(x - 36)$

   C $y - 36 = 2.2(x - 2)$

   D $y - 2.2 = 2(x - 36)$

6. The average height of a 5-year old girl is 44 inches, and the average growth rate per year is 2.4 inches. Write an equation in point-slope form that describes the height of a girl y based on her age x.

   F $y - 2.4 = 44(x - 5)$

   G $y - 44 = 2.4(x - 5)$

   H $y - 44 = 5(x - 2.4)$

   J $y - 5 = 2.4(x - 44)$

7. Write the equation from Exercise 6 in slope-intercept form.

   A $y = 2.4x - 100.6$

   B $y = 44x - 217.6$

   C $y = 5x + 32$

   D $y = 2.4x + 32$

8. Use the equation in Exercise 6 to find the average height of a 13-year old girl.

   F 56.3 in.

   G 63.2 in.

   H 69.4 in.

   J 97 in.

Name ______________________ Date __________ Class __________

LESSON 12-4

Reading Strategies

Use a Procedure

To find the slope of a line, you can use the coordinates for two points on the line.

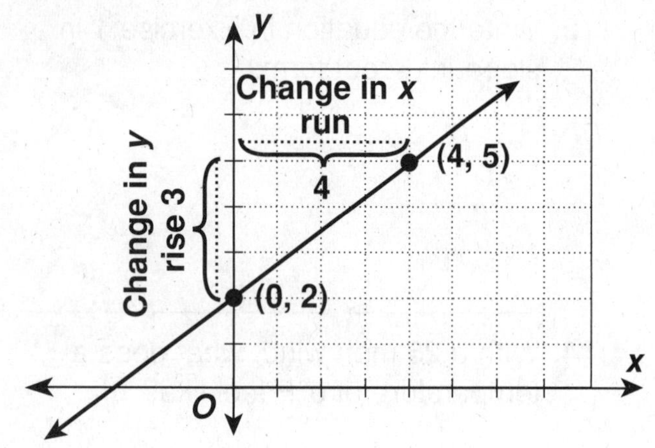

Step 1: Subtract to find the difference between the *y*-coordinates of the two points: $5 - 2 = 3$.

Step 2: Subtract to find the difference between the *x*-coordinates of the two points: $4 - 0 = 4$.

Step 3: Write the ratio of the differences. The difference between the *y*-coordinates is 3. The difference between the *x*-coordinates is 4. The slope is the ratio $\frac{3}{4}$.

When you have the slope of a line and a point it passes through, you can write an equation for the line in **point-slope form.**

Point on the line	Slope	Point-slope form
(x_1, y_1)	m	$y - y_1 = m(x - x_1)$
$(4, 5)$	$\frac{3}{4}$	$y - 5 = \frac{3}{4}(x - 4)$

Use the above example to answer each question.

1. What ratio is shown by the slope of a line?

____

2. How can you find the slope of a line?

____

____

____

3. What information do you need to write an equation for a line in point-slope form?

____

Name ______________________ Date __________ Class __________

LESSON 12-4

Puzzles, Twisters & Teasers

Get a Clue!

Identify a point on each line and the slope of the line. Then use the slope values to answer the riddle.

1. $y + 1 = \frac{2}{3}(x + 7)$ point = ________ slope = ________ **S**
2. $y + 1 = 11(x - 1)$ point = ________ slope = ________ **C**
3. $y - 2 = -\frac{1}{6}(x - 11)$ point = ________ slope = ________ **N**
4. $y + 7 = 1(x - 5)$ point = ________ slope = ________ **L**
5. $y + 7 = 3(x + 4)$ point = ________ slope = ________ **E**
6. $y - 9 = 5(x - 12)$ point = ________ slope = ________ **B**
7. $y - 11 = 14(x - 8)$ point = ________ slope = ________ **H**
8. $y - 4 = -2(x + 7)$ point = ________ slope = ________ **O**
9. $y - 3 = -1.8(x - 5.6)$ point = ________ slope = ________ **R**
10. $y + 8 = -6(x - 9)$ point = ________ slope = ________ **K**

What do you call a dog detective?

___	___	___	___	___	___	___	___
$\frac{2}{3}$	14	3	−1.8	1	−2	11	−6

___	___	___	___	___
5	−2	$-\frac{1}{6}$	3	$\frac{2}{3}$

Name ______________________ Date __________ Class __________

LESSON 12-5

Practice A

Direct Variation

The following tables show direct variation for the given equation. Complete the missing information in the tables.

1. $y = 2x$

x	−10	−7		3			15	22
y			−8		12	24		

2. $y = \frac{1}{3}x$

x	−21		−9	3	8		19	
y		−5				4		10

3. Make a graph to determine whether the data sets show direct variation.

x	y
−8	−4
−6	−3
0	0
2	1
4	2
6	3

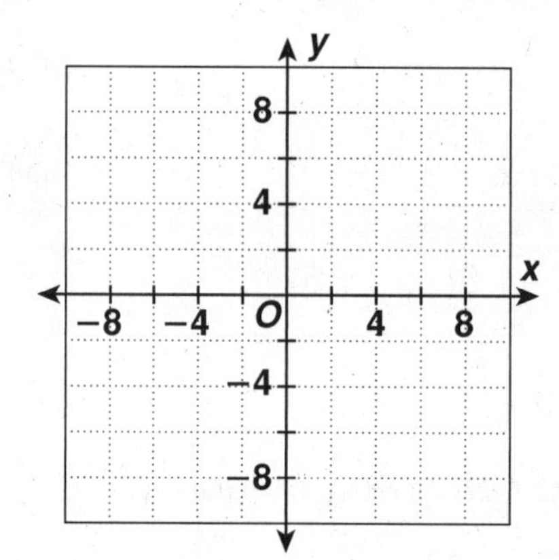

____

Find each equation of direct variation, given that *y* varies directly with *x*.

4. y is 10 when x is 2. __________

5. y is 42 when x is −6. __________

6. y is −50 when x is 5. __________

7. y is 15 when x is 30. __________

8. At a constant speed, the gasoline a car uses varies directly with the distance the car travels. A car uses 10 gallons of gasoline to travel 210 miles. How many gallons will the car use to travel 294 miles? __________

Name ______________________ Date ____________ Class ____________

Practice B
Direct Variation

Make a graph to determine whether the data sets show direct variation.

1.

x	y
6	9
4	6
0	0
−2	−3
−8	−12

____

2. Write the equation of direct variation for Exercise 1.

____

Find each equation of direct variation, given that y varies with x.

3. y is 32 when x is 4

____

4. y is −10 when x is −20

____

5. y is 63 when x is −7

____

6. y is 40 when x is 50

____

7. y is 87.5 when x is 25

____

8. y is 90 when x is 270

____

9. The table shows the length and width of various U.S. flags. Determine whether there is direct variation between the two data sets. If so, find the equation of direct variation.

Length (ft)	2.85	5.7	7.6	9.88	11.4
Width (ft)	1.5	3	4	5.2	6

____

____

____

Name ____________________ Date __________ Class __________

LESSON 12-5

Practice C

Direct Variation

Find each equation of direct variation, given that y varies directly with x.

1. y is 189 when x is 45

____

2. y is 456 when x is 3800

____

3. y is 763 when x is 981

____

4. y is $171\frac{3}{4}$ when x is 916

____

Tell whether each equation represents direct variation between x and y.

5. $y = \frac{9}{10}x$

____

6. $y = xy - 8$

____

7. $-5x - y = 0$

____

8. $y = \frac{24}{x}$

____

9. $\frac{y}{x} = 8.25$

____

10. $x - y = -10$

____

11. $x = y$

____

12. $\frac{1}{3}y = x$

____

13. The following table shows the distance on a map in inches x and the actual distance between two cities in miles, y. Determine whether there is direct variation between the two data sets. If so, find the equation of direct variation.

x	$456\frac{1}{4}$	$3\frac{1}{2}$	4	5	$7\frac{1}{4}$	8	$9\frac{1}{8}$	11
y	75	175	200	350	$362\frac{1}{2}$	400	$456\frac{1}{4}$	550

____

14. A person's weight on Earth varies directly with a person's estimated weight on Venus. If a person weighs 110 pounds on Earth, he or she would weigh an estimated 99.7 pounds on Venus. If a person weighs 125 pounds on Earth, what would be his or her estimated weight to the nearest tenth of a pound on Venus?

____

Name ______________________ Date __________ Class __________

LESSON **12-5**

Reteach

Direct Variation

Two data sets have **direct variation** if they are related by a constant ratio, the **constant of proportionality.** A graph of the data sets is linear and passes through (0, 0).

$y = kx$ **equation of direct variation,** where k is the constant ratio

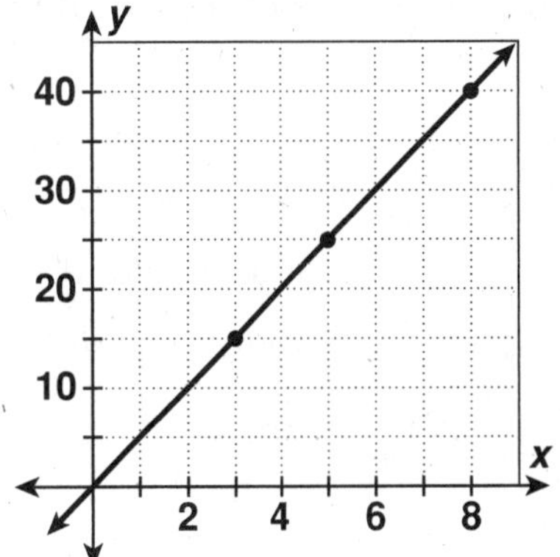

To determine whether two data sets have direct variation, you can compare ratios. You can also graph the data sets on a coordinate grid.

x	3	5	8
y	15	25	40

$\frac{y}{x} = \frac{15}{3} = \frac{25}{5} = \frac{40}{8} = \frac{5}{1}$ ← constant ratio

$k = 5 \rightarrow y = 5x$

The graph of the data sets is linear and passes through (0, 0).

So, the data sets show direct variation.

Determine whether the data sets show direct variation. If there is a constant ratio, identify it and write the equation of direct variation. Plot the points and tell whether the graph is linear.

1. 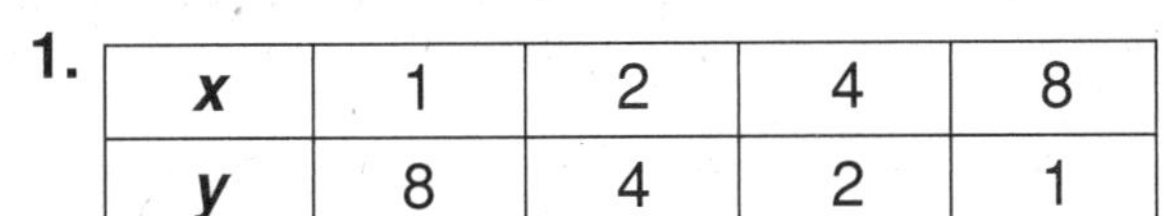

x	1	2	4	8
y	8	4	2	1

constant ratio? ______________

If yes, equation. ______________

Is the graph linear? ______________

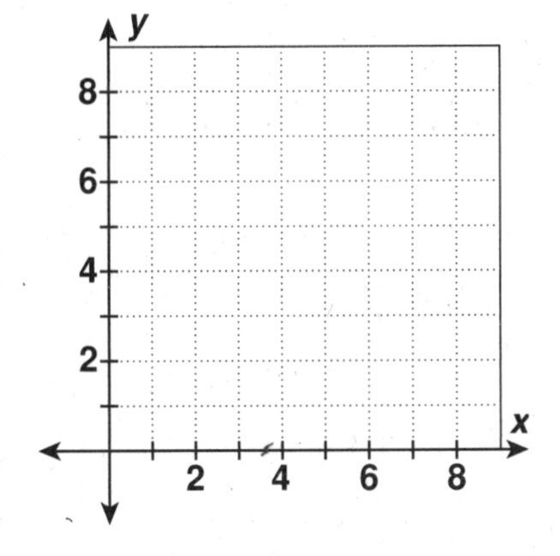

2. 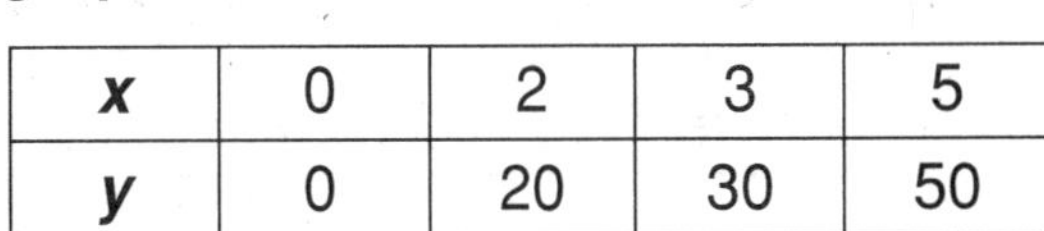

x	0	2	3	5
y	0	20	30	50

constant ratio? ______________

If yes, equation. ______________

Is the graph linear? ______________

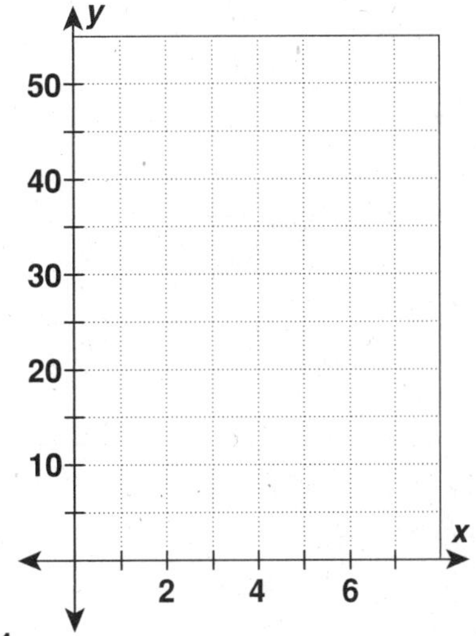

Name ______________________ Date __________ Class __________

LESSON 12-5

Challenge

Different Paths, Same Result

Problems of direct variation can be solved with two methods.
If r varies directly with h, and $r = 13.5$ when $h = 3$, find r when $h = 7$.

Method 1: Find the constant of variation.

$\frac{r}{h} = k$

$\frac{13.5}{3} = k$ Use a pair of known values.

$4.5 = k$ constant of variation

$r = 4.5h$ equation of variation

$r = 4.5(7) = 31.5$

So, when $h = 7$, $r = 31.5$.

Method 2: Write a proportion.

$\frac{r_1}{h_1} = \frac{r_2}{h_2}$

$\frac{13.5}{3} = \frac{r_2}{7}$ Use all known values.

$3r_2 = 13.5(7)$ Cross multiply.

$\frac{3r_2}{3} = \frac{94.5}{3}$

$r_2 = 31.5$

Use both methods to solve each problem.

1. y varies directly as x. If $y = 16$ when $x = 5$, find y when $x = 9$.

______________________ ______________________

______________________ ______________________

______________________ ______________________

______________________ ______________________

______________________ ______________________

So, when $x = 9$, $y =$ ________.

2. A varies directly as s^2. If $A = 75$ when $s = 5$, find A when $s = 7$.

______________________ ______________________

______________________ ______________________

______________________ ______________________

______________________ ______________________

So, when $s = 7$, $A =$ ________

Name ______________________ Date ____________ Class ____________

LESSON 12-5

Problem Solving

Direct Variation

Determine whether the data sets show direct variation. If so, find the equation of direct variation.

1. The table shows the distance in feet traveled by a falling object in certain times.

Time (s)	0	0.5	1	1.5	2	2.5	3
Distance (ft)	0	4	16	36	64	100	144

____

____

2. The R-value of insulation gives the material's resistance to heat flow. The table shows the R-value for different thicknesses of fiberglass insulation.

Thickness (in)	1	2	3	4	5	6
R-value	3.14	6.28	9.42	12.56	15.7	18.84

____

____

3. The table shows the lifting power of hot air.

Hot Air (ft^3)	50	100	500	1000	2000	3000
Lift (lb)	1	2	10	20	40	60

____

____

4. The table shows the relationship between degrees Celsius and degrees Fahrenheit.

° **Celsius**	−10	−5	0	5	10	20	30
° **Fahrenheit**	14	23	32	41	50	68	86

____

____

The relationship between your weight on Earth and your weight on other planets is direct variation. The table below shows how much a person who weights 100 lb on Earth would weigh on the moon and different planets.

Solar System Objects	**Weight** (lb)
Moon	16.6
Jupiter	236.4
Pluto	6.7

5. Find the equation of direct variation for the weight on earth e and on the moon m.

A $m = 0.166e$ **C** $m = 6.02e$

B $m = 16.6e$ **D** $m = 1660e$

6. How much would a 150 lb person weigh on Jupiter?

F 63.5 lb **H** 354.6 lb

G 286.4 lb **J** 483.7 lb

7. How much would a 150 lb person weigh on Pluto?

A 5.8 lb **C** 12.3 lb

B 10.05 lb **D** 2238.8 lb

Name ______________________ Date ____________ Class ____________

LESSON 12-5

Reading Strategies

Use Tables and Graphs

When quantities are related proportionally by a constant multiplier, they have **direct variation.**

This table shows the relationship between the number of glasses filled and the amount of juice needed to fill them. The amount of juice needed *varies directly* with the number of glasses filled.

Glasses	1	2	3	4
Juice Needed	8 oz	16 oz	24 oz	32 oz

1. What are the quantities that form this direct variation?

____

2. What is the constant multiplier?

____

A graph of a direct variation is always linear and always passes through (0,0).

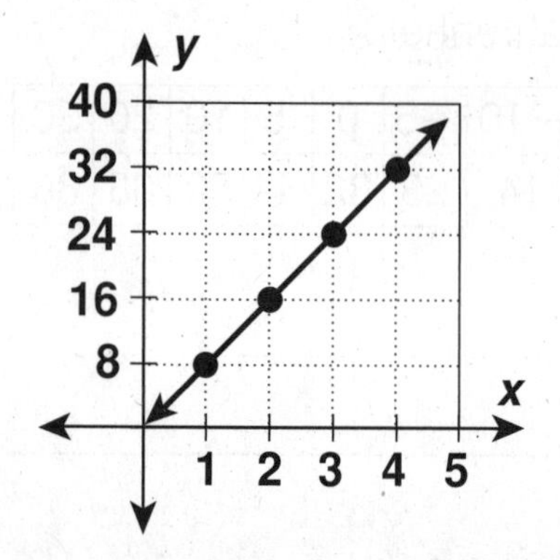

3. What do the *x*-values on the graph stand for?

____

4. What do the *y*-values on the graph stand for?

____

5. What does the ordered pair (2, 16) mean?

____

6. Write an ordered pair for 3 glasses and the amount of juice needed.

____

Name ______________________ Date ____________ Class ____________

Puzzles, Twisters & Teasers

It Just Doesn't Hold Water!

Circle words from the list in the word search (horizontally, vertically or diagonally). You will also find a word that answers the riddle.

direct	variation	constant	proportionality	ratio
graph	quantity	algebra	table	data

P R O P O R T I O N A L I T Y
V A C V B N A I K L L Q W E R
A T D F G H B U J M G R A P H
R I E R T Y L T H N E Q T Y U
I O B G H U E W D V B U I O P
A A S D F G H J K L R A A S D
T S I E V E X D A T A N F G H
I Z X C V B N M K I J T K L Q
O D I R E C T H U J I I Z X C
N C O N S T A N T D F T V B N
L K J H G F D S A T B Y M K I

What is as round as a dishpan and as deep as a tub, yet the ocean could not fill it?

A ____________________.

Name ______________________ Date ____________ Class ____________

LESSON 12-6

Practice A

Graphing Inequalities in Two Variables

1. The graph shows $y = x + 2$. Shade the the inequality $y \leq x + 2$.

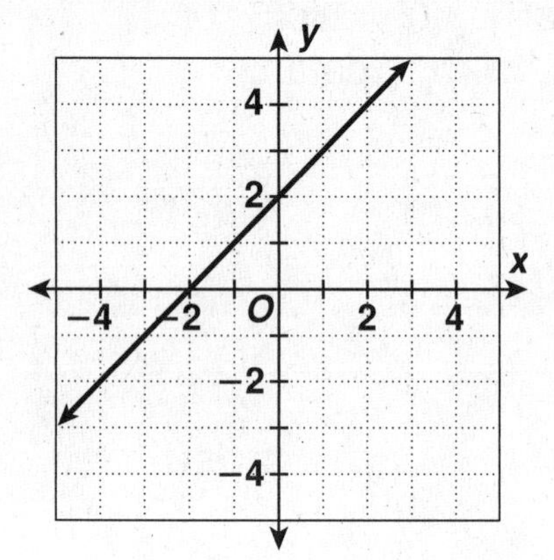

2. The graph shows $y = x$. Shade the side of the line to show the inequality $y \geq x$.

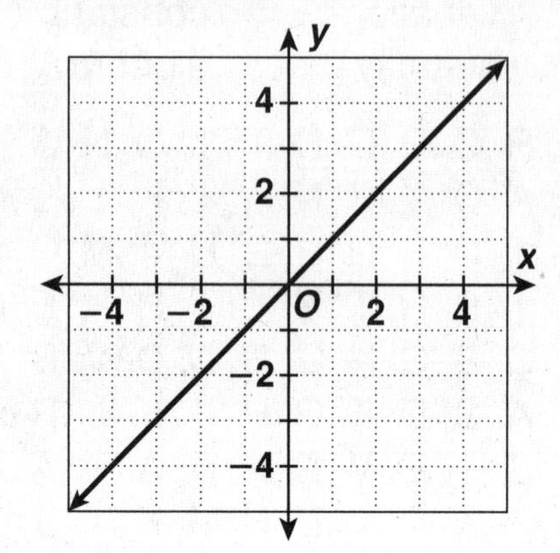

Solve each inequality for *y*.

3. $x + y \geq 8$ ______________

4. $3x - y > 6$ ______________

5. $x - y \geq -2$ ______________

6. $4x + y \leq -3$ ______________

Graph each inequality.

7. $y \leq x - 1$

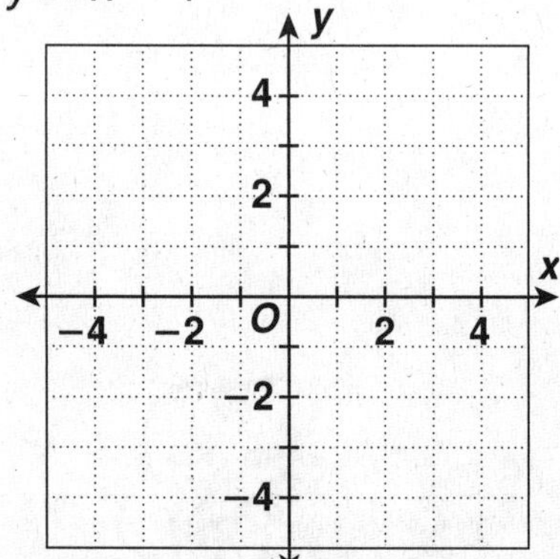

8. $y \geq 2x + 1$

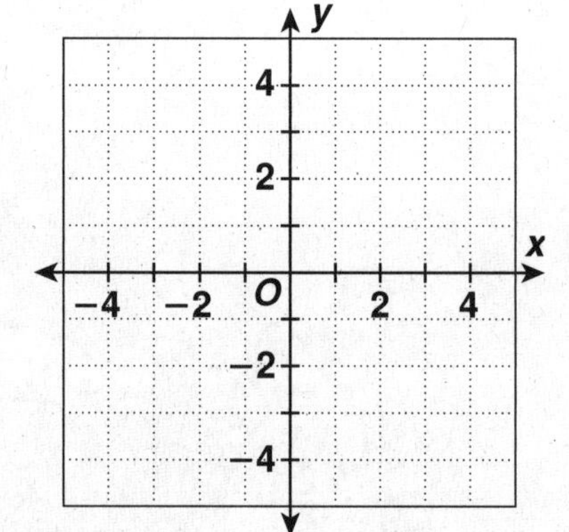

9. Georgia has scored at least 12 points in each of her basketball games this year. She has scored both 2-point and 3-point field goals in every game. Let x equal the 2-point shots and y equal the 3-point shots. Write and graph an inequality to show the number of points she scored each game.

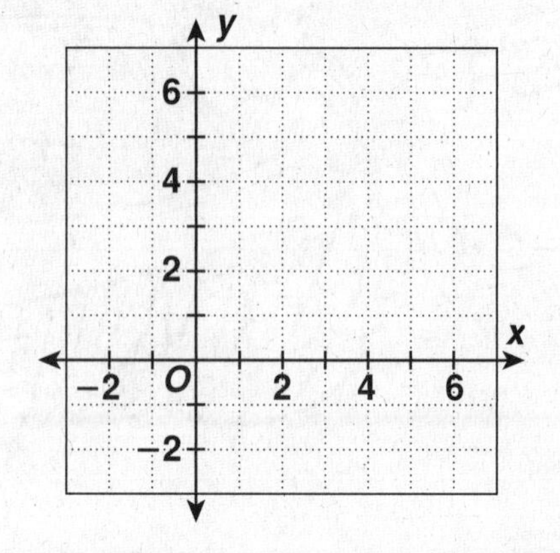

____

Name ______________________ Date __________ Class __________

LESSON **12-6**

Practice B

Graphing Inequalities in Two Variables

Graph each inequality.

1. $y \geq 2x + 3$

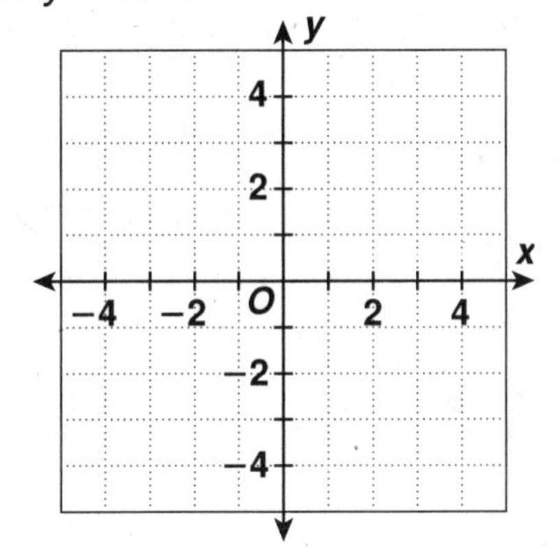

2. $y - 4x \leq 1$

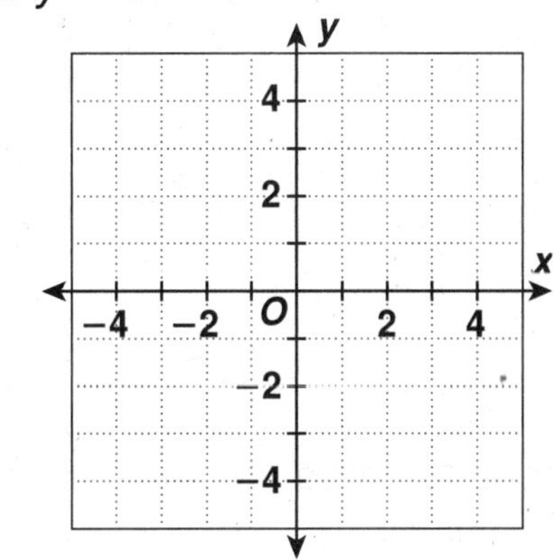

3. $2(3x - y) > 6$

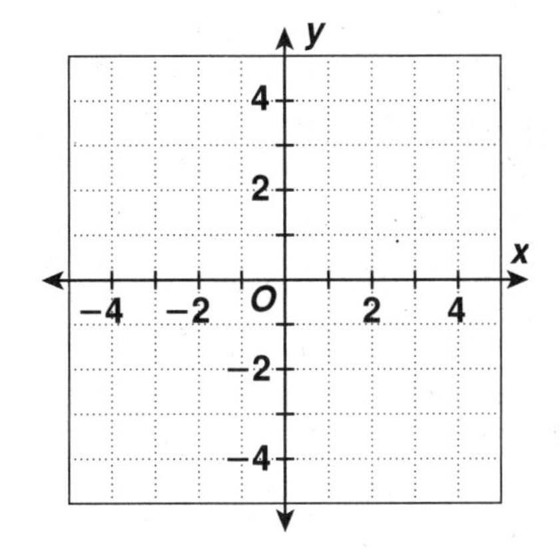

4. $y \geq \frac{3}{4}x - 1$

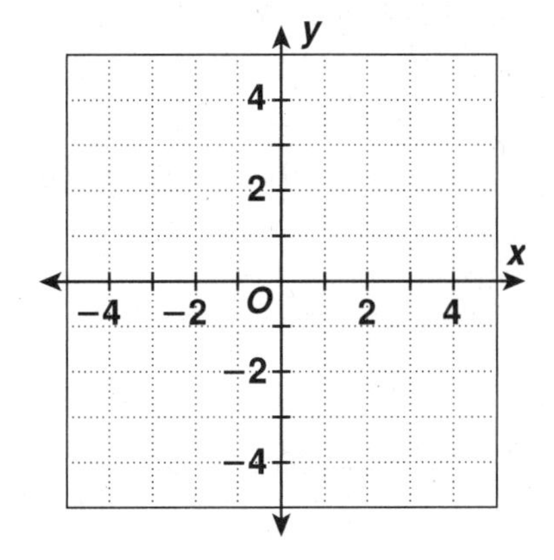

5. a. A theater club hopes to raise at least $550 on the opening night of its new show. Student tickets for the show cost $2.75, and adult tickets cost $5.50. Write and graph an inequality showing the numbers of tickets that would meet the club's goal.

____

____

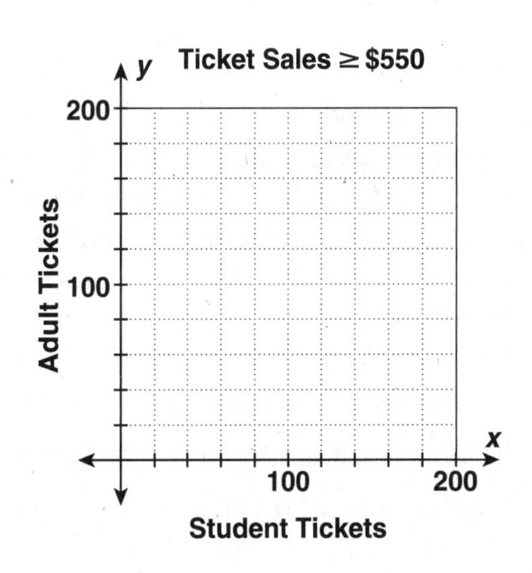

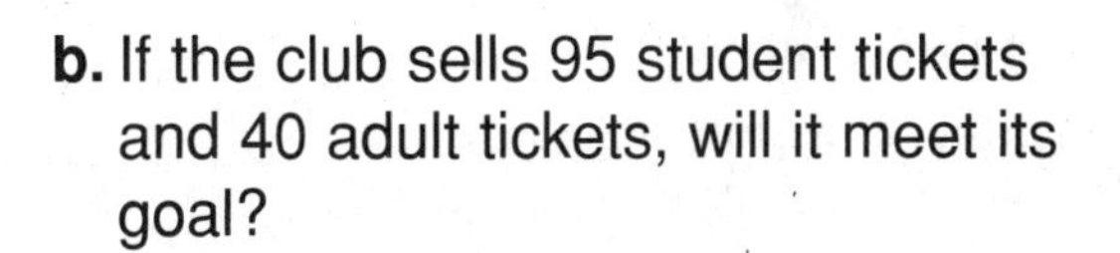

b. If the club sells 95 student tickets and 40 adult tickets, will it meet its goal?

____

Name ______________________ Date __________ Class __________

LESSON 12-6
Practice C
Graphing Inequalities in Two Variables

Tell whether the given ordered pair is a solution of each inequality shown.

1. $2x + 3y \geq 6$, (2, 1) ______

2. $3x - 5y > -10$, (4, 7) ______

3. $x - 2y \geq 8$, (5, −2) ______

4. $5x - 4y \geq -9$, (−3, −2) ______

5. $y > \frac{1}{2}x + \frac{3}{4}$, (−2, 1) ______

6. $y \geq \frac{2}{3}x - 8$, (6, −7) ______

7. a. Graph the inequality $3x - 2y \geq 4$.

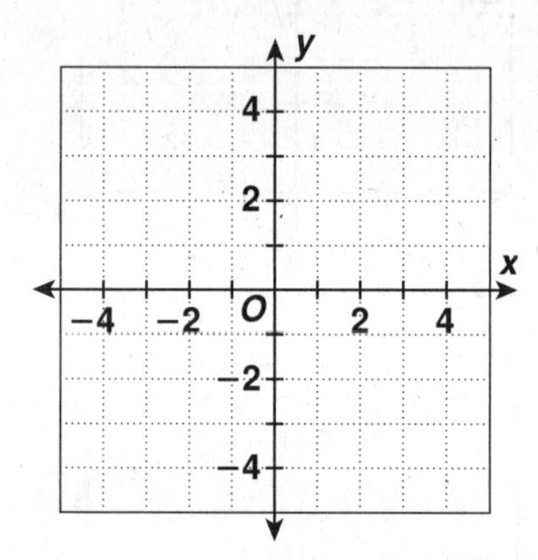

b. Name an ordered pair that is a solution of the inequality.

____

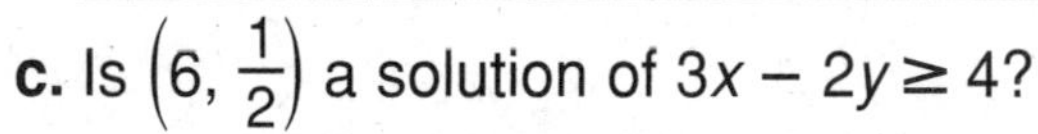

c. Is $\left(6, \frac{1}{2}\right)$ a solution of $3x - 2y \geq 4$?

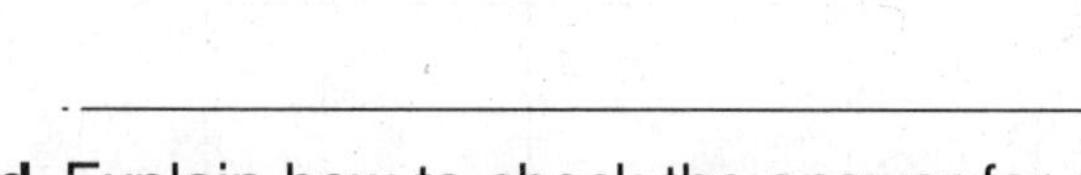

____

d. Explain how to check the answer for c.

____

____

e. Which side of the line $3x - 2y = 4$ is shaded?

____

f. Name an ordered pair that is a solution of $3x - 2y \leq 4$.

____

8. Jaime is having a party and has budgeted \$120 for the party. He plans on serving pizzas that cost \$6 each and soft drinks that cost \$0.75 each. Let *x* represent the number of pizzas and *y* represent the number of soft drinks. Write an inequality showing the numbers of pizzas and soft drinks he could buy while staying within his budget.

____

9. If Jaime invites 25 people and plans on each person having 2 soft drinks and $\frac{1}{2}$ pizza, will he have budgeted enough money? ______

Name ______________________ Date __________ Class __________

LESSON 12-6

Reteach

Graphing Inequalities in Two Variables

A **boundary line** divides the coordinate plane into two *half-planes*.

When $y = -x + 4$ is a boundary line:
All the points on the line satisfy the equation.
(5, −1) is on the line since $-1 = -5 + 4$.
All the points in the half-plane above the line satisfy the linear inequality $y > -x + 4$.
(5, 3) is in the half-plane above the line since $3 > -5 + 4$.
All the points in the half-plane below the line satisfy the linear inequality $y < -x + 4$.
(−2, 1) lies in the half-plane below the line since $1 < -(-2) + 4$.

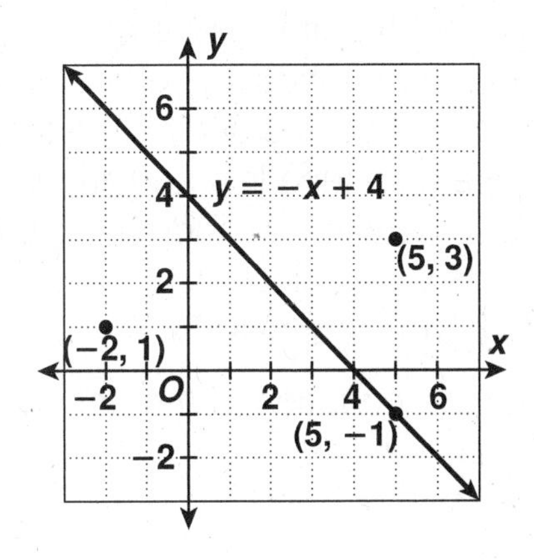

When a boundary line is in the form $y = mx + b$, points in the half-plane above the line satisfy the inequality $y > mx + b$ and points in the half-plane below the line satisfy the inequality $y < mx + b$.

Complete the linear inequality that each point satisfies.

1. (1, −2) is in the half-plane below the boundary line $y = 3x - 4$.

   The boundary line is in the form $y = mx + b$; so, (1, −2) satisfies the linear inequality y ____ $3x - 4$.

2. (−3, 7) is in the half-plane below the boundary line $y = -2x + 6$.

   The boundary line is in the form $y = mx + b$; so, (−3, 7) satisfies the linear inequality y ____ $-2x + 6$.

Write the linear inequality whose solution set is shaded on each graph. The dashed boundary line is not included in the solution set.

3.

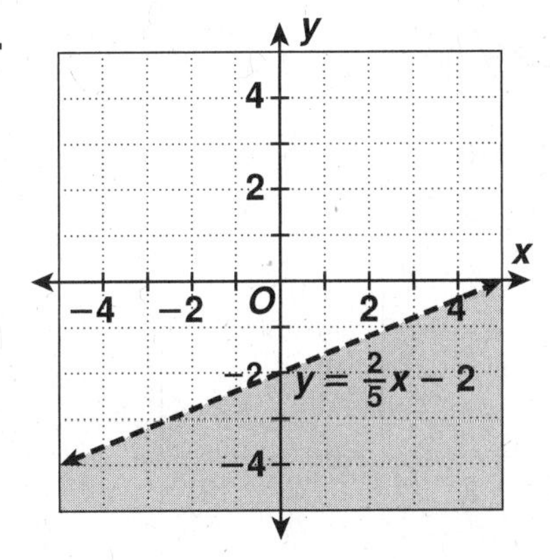

____

4.

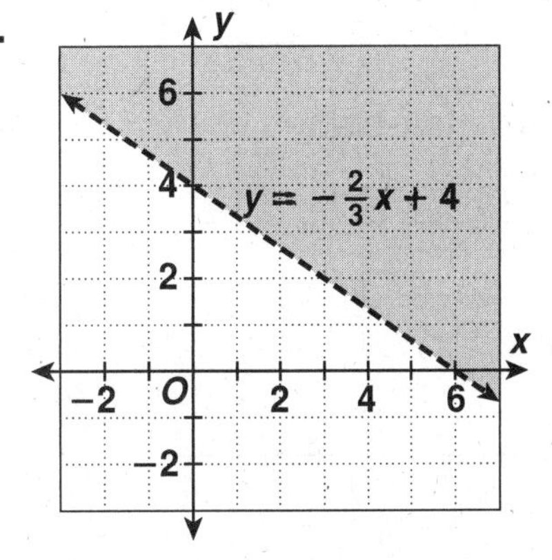

____

Name ______________________________ Date ____________ Class ____________

LESSON 12-6

Reteach

Graphing Inequalities in Two Variables (continued)

To graph the solution set of a linear inequality:

Write the equation of the boundary line in the form $y = mx + b$.
Graph the boundary line.
Use a dashed line and shade the region above if $y > mx + b$.
Use a dashed line and shade the region below if $y < mx + b$.
Use a solid line and shade the region above if $y \geq mx + b$.
Use a solid line and shade the region below if $y \leq mx + b$.

Graph the inequality $2x + y \leq -6$:

Write the equation of the boundary line in $y = mx + b$ form.

$$2x + y = -6$$
$$-2x \qquad -2x$$
$$y = -2x - 6$$

slope $= -2$, y-intercept $= -6$

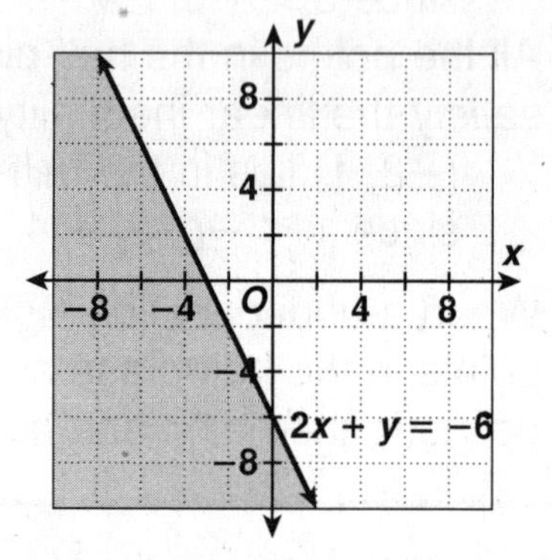

Since the inequality uses $\leq$, draw a solid boundary line and shade the half-plane below.

Graph the solution set of each inequality.

5. $3x + y \geq 4$

Rewrite equation $3x + y = 4$.

$y =$ ____________

slope = ______, y-intercept = ______

Given symbol is $\geq$, so draw

boundary line ____________

and shade half-plane ____________.

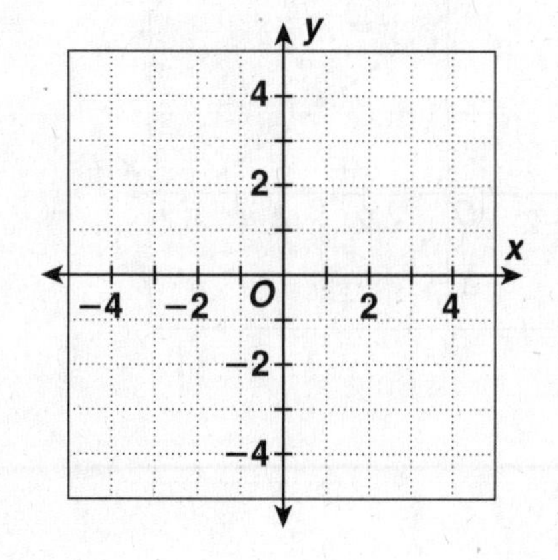

6. $2y - x < 6$

Rewrite equation $2y - x = 6$.

$y =$ ____________

slope = ______, y-intercept = ______

Given symbol is $<$, so draw

boundary line ____________

and shade half-plane ____________.

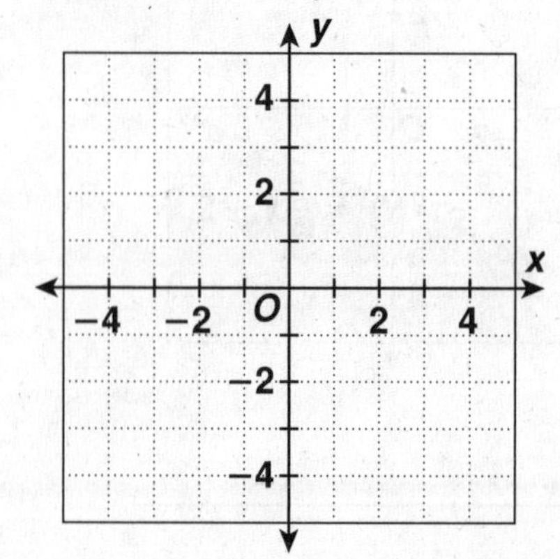

Name ________________________________ Date ___________ Class ___________

LESSON 12-6

Challenge

Two at a Time

To solve a system of two linear inequalities graphically find the solution set for each linear inequality and mark off the part that overlaps.

Graph the solution set of the system: $x + y \geq 1$
$y < x - 3$

Work with the first inequality.
Rewrite $x + y = 1$ as $y = -x + 1$.
Since line is now in $y = mx + b$ form and given symbol is $\geq$, draw solid boundary line and shade half-plane above line.

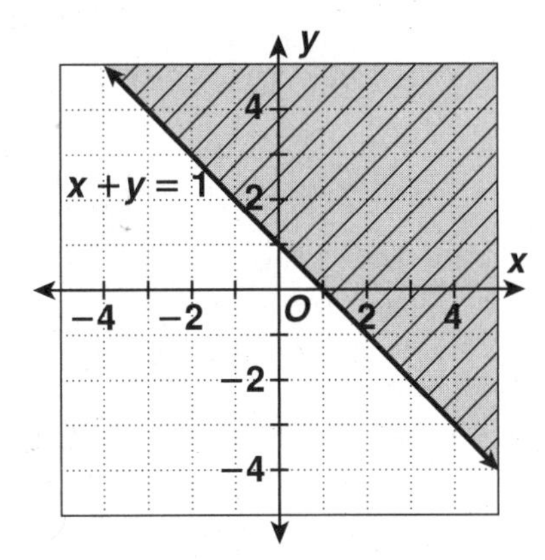

Work with the second inequality.
Since boundary line $y = x - 3$ is in $y = mx + b$ form and given symbol is $<$, draw dashed boundary line and shade half-plane below the line.

Use a shading opposite to the first shading so that overlap is visible.

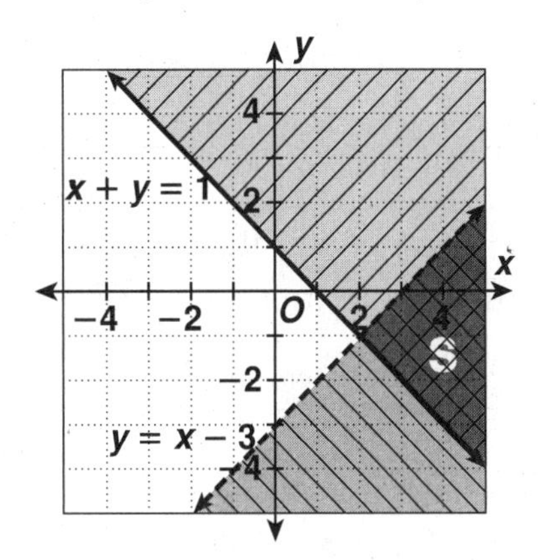

So, the solution set to the system is the cross-hatched region labeled *S*.

Graph the solution set *S* for each system.

1. $y \geq 3x + 1$
$y < x + 1$

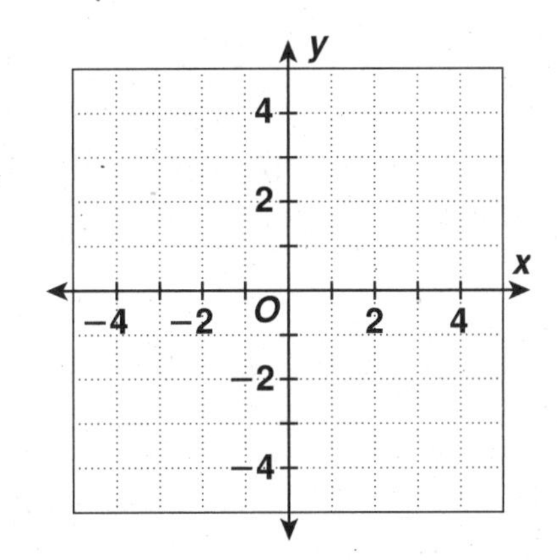

2. $2x - y \leq 4$
$3x + y < 6$

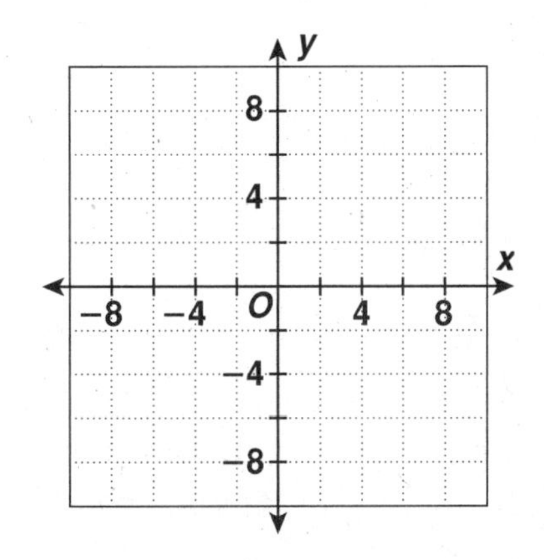

Name ______________________ Date __________ Class __________

LESSON 12-6

Problem Solving

Graphing Inequalities in Two Variables

The senior class is raising money by selling popcorn and soft drinks. They make \$0.25 profit on each soft drink sold, and \$0.50 on each bag of popcorn. Their goal is to make at least \$500.

1. Write an inequality showing the relationship between the sales of x soft drinks and y bags of popcorn and the profit goal.

____

2. Graph the inequality from exercise 1.

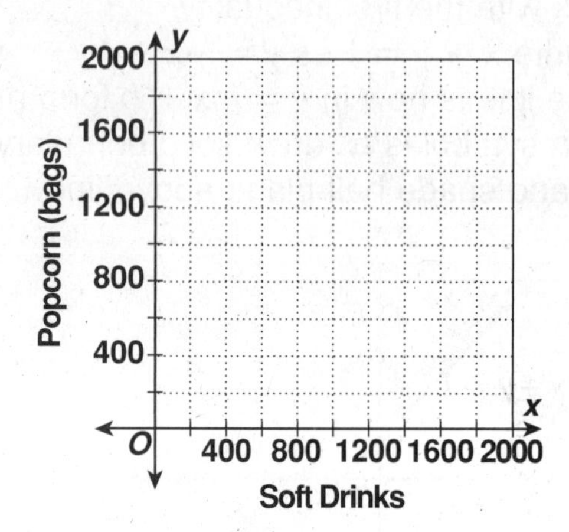

3. List three ordered pairs that represent a profit of exactly \$500.

____

____

4. List three ordered pairs that represent a profit of more than \$500.

____

____

5. List three ordered pairs that represent a profit of less than \$500.

____

____

A vehicle is rated to get 19 mpg in the city and 25 mpg on the highway. The vehicle has a 15-gallon gas tank. The graph below shows the number of miles you can drive using no more than 15 gallons.

6. Write the inequality represented by the graph.

A $\frac{x}{19} + \frac{y}{25} < 15$

B $\frac{x}{19} + \frac{y}{25} \leq 15$

C $\frac{x}{19} + \frac{y}{25} \geq 15$

D $\frac{x}{19} + \frac{y}{25} > 15$

7. Which ordered pair represents city and highway miles that you can drive on one tank of gas?

F (200, 150) H (250, 75)

G (50, 350) J (100, 175)

8. Which ordered pair represents city and highway miles that you cannot drive on one tank of gas?

A (100, 200) C (50, 275)

B (150, 200) D (250, 25)

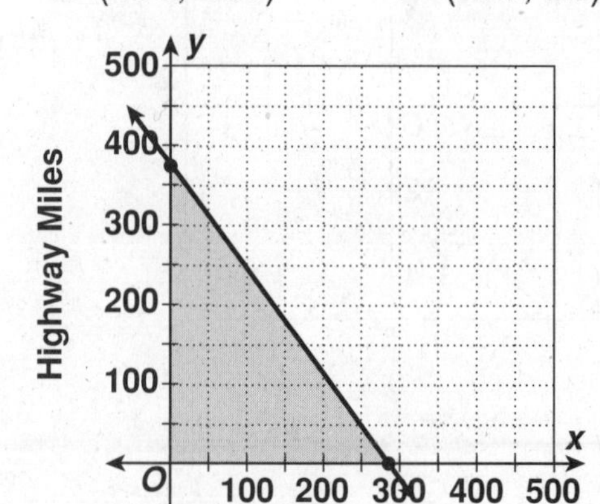

Name ______________________ Date ____________ Class ____________

LESSON 12-6

Reading Strategies

Use a Graphic Aid

The graph of a **linear inequality** is half the coordinate plane. The boundary line is the graph of a linear equation.

Linear inequality → $x + y < 10$

Boundry Line → $x + y = 10$

Follow these steps to graph the linear inequality $x + y < 10$.

Step 1: Graph the linear equation showing the boundary line.

x	y
3	7
5	5

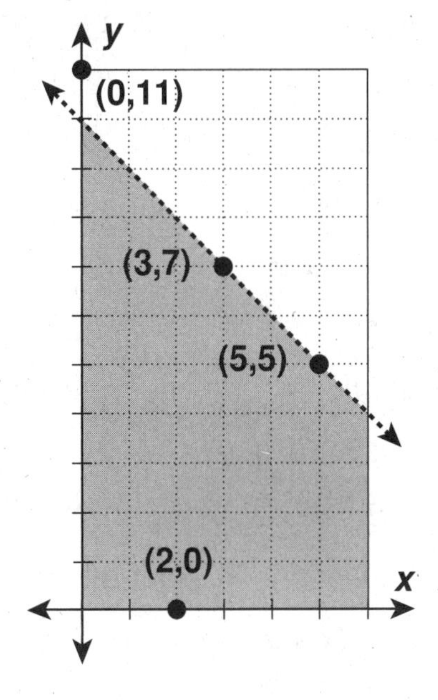

Draw a dashed line because none of the points on the line are part of the solution.

Step 2: Test a point on each side of the line.

(2, 0)	(0, 11)
$2 + 0 < 10$	$0 + 11 < 10$
$2 < 10$ *true*	$11 < 10$ *not true*

Step 3: Shade the side of the line that contains (2, 0).

For 1–4, refer to the example above.

1. What is the first step in graphing a linear inequality?

____

2. Why is the line dashed instead of solid?

____

3. What is the second step in graphing a linear inequality?

____

4. What does the shaded region of the graph show?

____

Name ________________________________ Date ____________ Class ____________

LESSON 12-6

Puzzles, Twisters & Teasers

Dog-Food for Thought!

Determine whether the given ordered pair is a solution of the inequality. Circle the letter of the correct answer. Then use the letters above the answers to solve the riddle.

1. $y > x + 1$ (0, 0)
   X solution — D not a solution
2. $y \leq x + 1$ (2, 1)
   O solution — Q not a solution
3. $3y + 4x \leq 12$ (0, 0)
   G solution — Z not a solution
4. $y < 330x + 0$ (5, 0)
   B solution — N not a solution
5. $y \leq 2x + 4$ (−7, 5)
   Y solution — I not a solution
6. $y > -x + 12$ (−3, 9)
   D solution — S not a solution
7. $y > -6x + 1$ (2, −14)
   H solution — C not a solution
8. $y > 3x - 3$ (3, 3)
   G solution — U not a solution
9. $y < 3(x - 2)$ (5, 1)
   I solution — R not a solution
10. $y \leq 7(x - 3)$ (5, 14)
    T solution — P not a solution
11. $y < -1.5x + 2.5$ (−1, 6)
    L solution — S not a solution

What do you get if you cross a beagle with some bread dough?

____

Name ______________________ Date __________ Class __________

LESSON 12-7

Practice A
Lines of Best Fit

Find the mean of the *x*- and *y*-coordinates of each set of data.

1.

x	1	2	4	5	7	3	6	8
y	5	6	7	8	10	7	10	11

____

2.

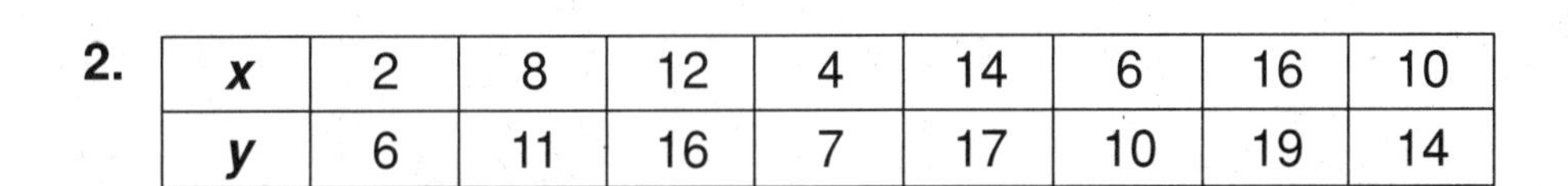

x	2	8	12	4	14	6	16	10
y	6	11	16	7	17	10	19	14

____

3. Plot the data points for Exercise 1.

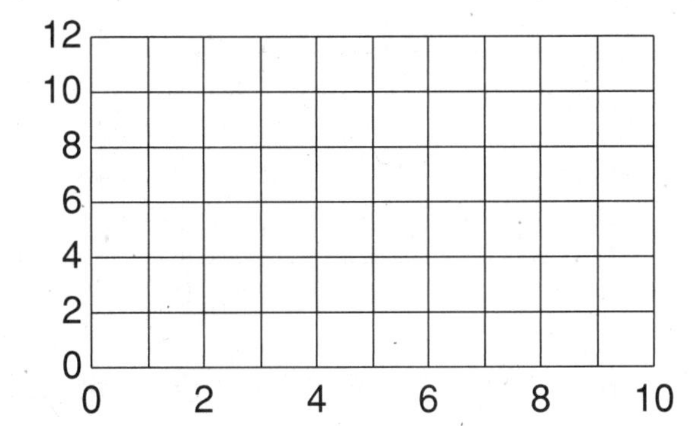

4. Plot the data points for Exercise 2.

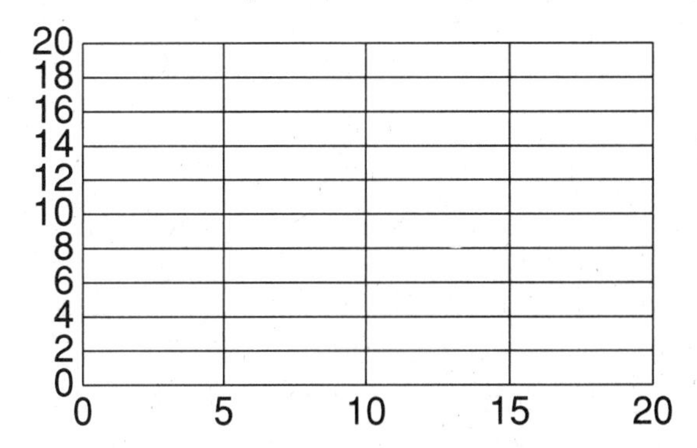

5. Draw a line through (x_m, y_m) of Exercise 3 to best fit the data.

____

6. Draw a line through (x_m, y_m) of Exercise 4 to best fit the data.

____

7. Write the equation of the line of best fit for Exercise 5.

____

8. Write the equation of the line of best fit for Exercise 6.

____

Name ______________________ Date ____________ Class ____________

LESSON 12-7

Practice B

Lines of Best Fit

Plot the data and find a line of best fit.

1.

x	20	30	50	60	80	90	110	120
y	13	20	40	54	75	82	100	112

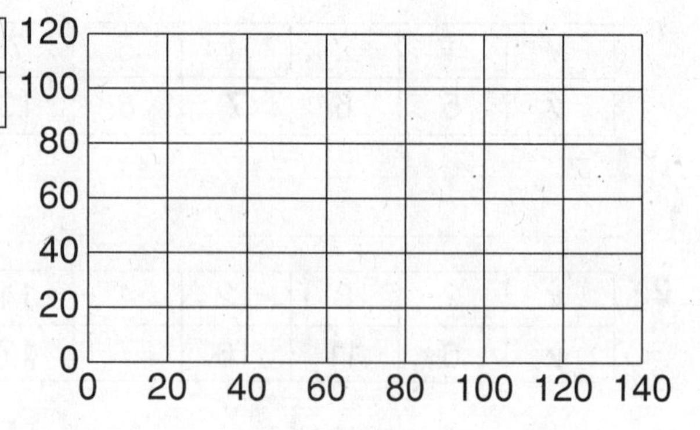

____

2.

x	1.9	2.9	4.8	2.5	3.9	2.3	6.3	3.4
y	26	34	58	31	52	27	76	48

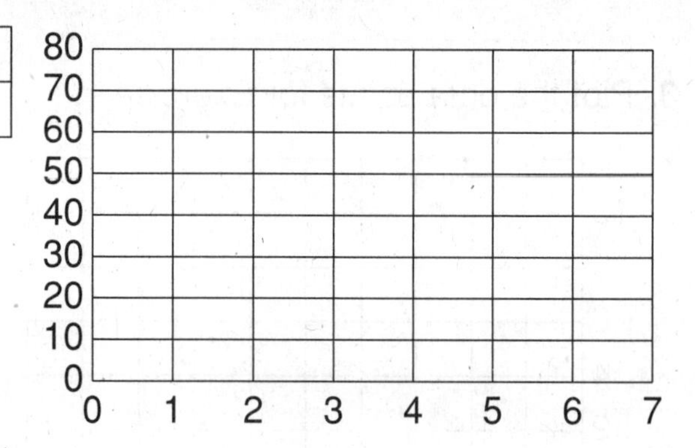

____

3. Find the line of best fit for the student enrollment data. Use the equation of the line to predict what the enrollment at Columbus Junior High School will be in year 10. Is it reasonable to make this prediction? Explain.

Enrollment	405	485	557	593	638	712
Year	1	2	3	4	5	6

____

____

____

____

____

____

____

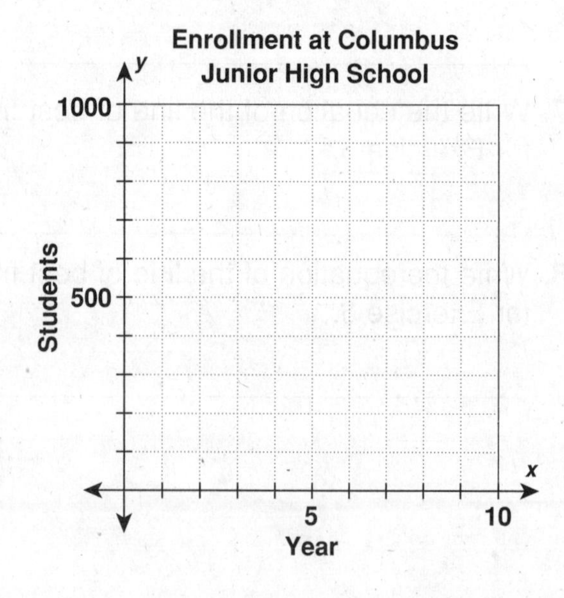

Name ______________________ Date __________ Class __________

LESSON 12-7

Practice C

Lines of Best Fit

Tell whether a line of best fit for each scatter plot would have a positive or negative slope. If a line of best fit would not be appropriate for the data, write *neither*.

1.

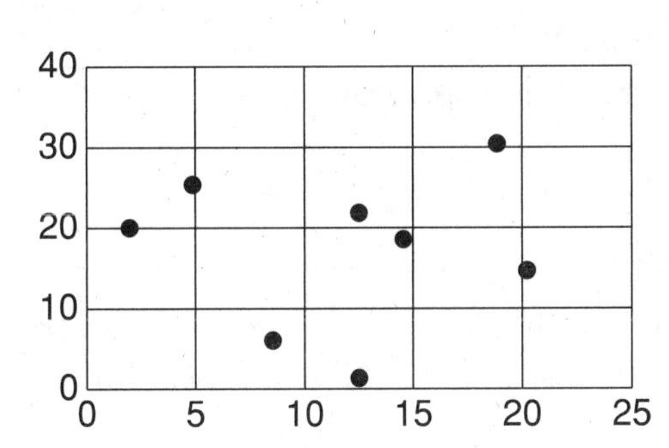

____

2.

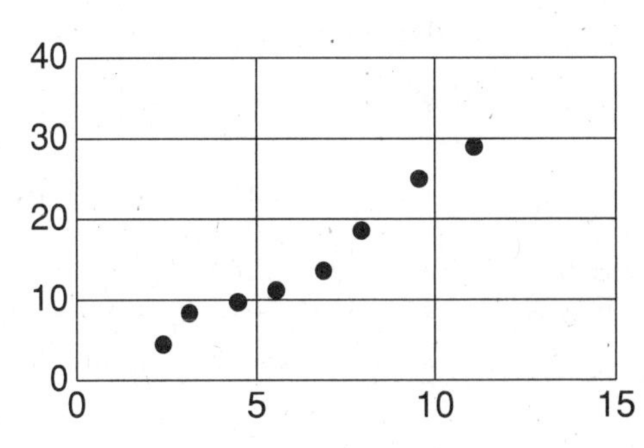

____

3.

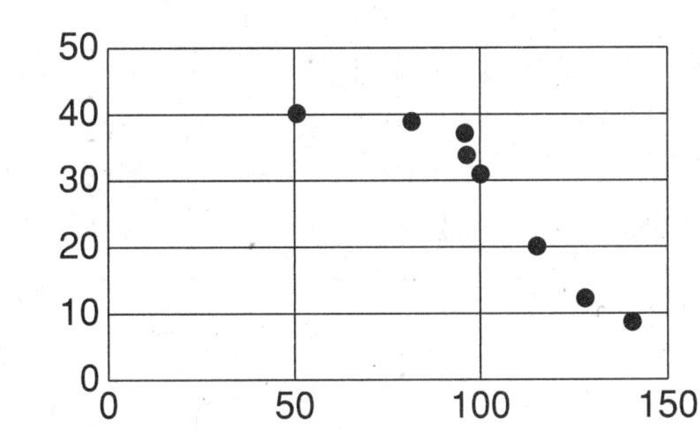

____

4.

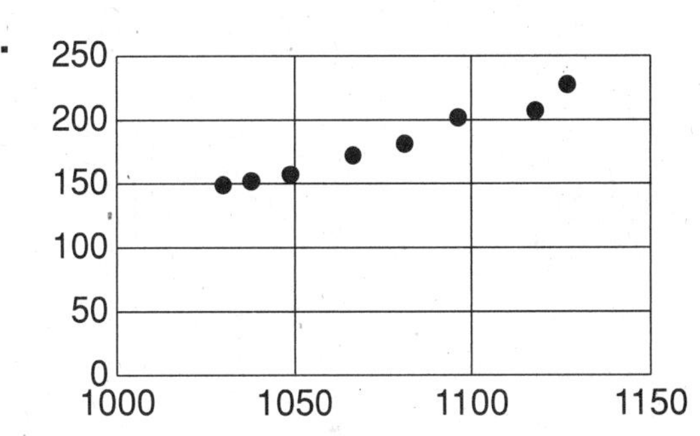

____

5. Mrs. Kani is a salesperson for a local manufacturer. She wants to ask for a raise based on her sales record for the last eight months. Find the line of best fit for her sales data and use the equation of the line to predict her sales for one month a year from now.

Sales ($)	7500	7740	7790	7900	8130	8200	8300	8480
Month	1	2	3	4	5	6	7	8

____

____

____

____

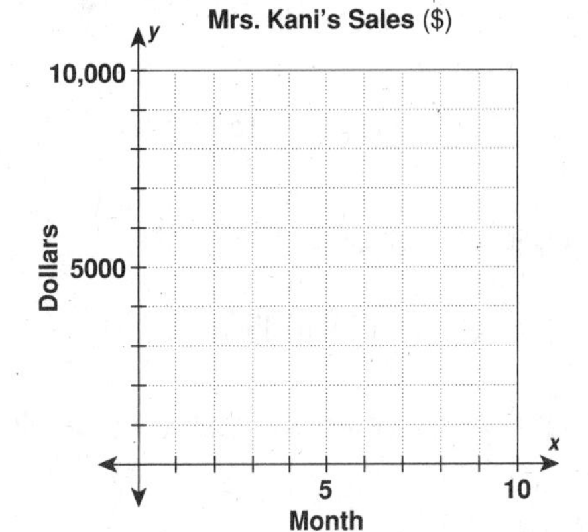

Name ______________________________ Date ____________ Class ____________

LESSON 12-7

Reteach

Lines of Best Fit

Line of best fit: a line drawn near the points on a scatter plot to show the trend between two sets of data

To draw a line of best fit:

Draw the line through as many points as you can.
Try to get an equal number of points above the line as below.
Ignore any outliers.
It may happen that none of the points lie on the line.

The line of best fit for the data in the scatter plot below slants up, indicating a positive correlation.

The slope of this line of best fit is positive and its *y*-intercept is about 3.

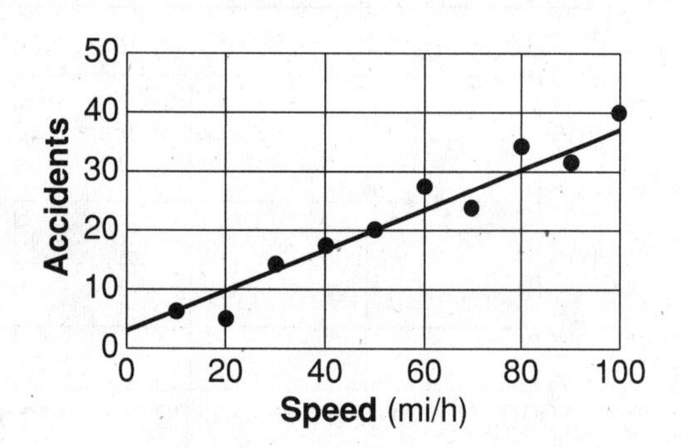

The line of best fit for the data in the scatter plot below slants down, indicating a negative correlation.

The slope of this line of best fit is positive and its *y*-intercept is about 28.

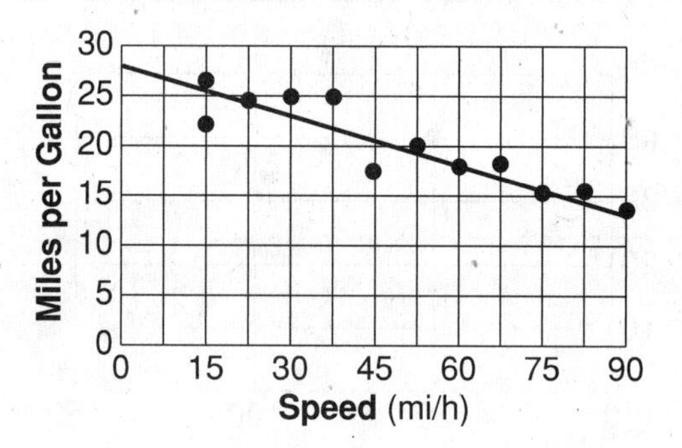

Draw a line of best fit for each graph. Describe the slope and find an approximate value for the *y*-intercept of the line of best fit.

1.

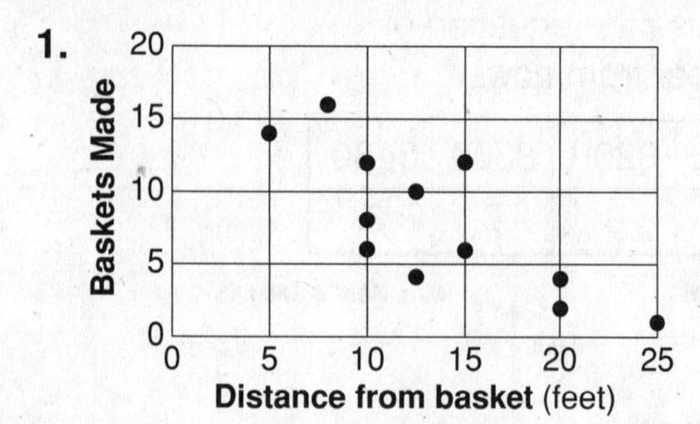

The line of best fit slants __________.

The slope of the line is __________.

The *y*-intercept is approximately ____.

2.

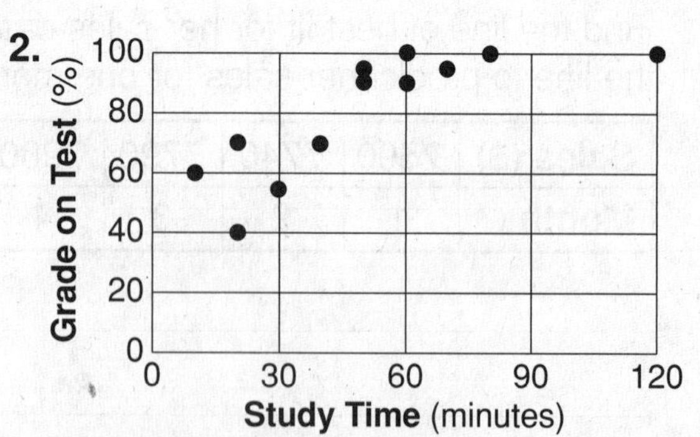

The line of best fit slants __________.

The slope of the line is __________.

The *y*-intercept is approximately ____.

Name ______________________ Date __________ Class __________

LESSON 12-7

Reteach

Lines of Best Fit (continued)

To write an equation for a line of best fit, you can use the slope-intercept form, $y = mx + b$.

After you have drawn the line of best fit, estimate its slope from any two points on the line whose coordinates you can read.

Draw a right triangle with the hypotenuse on the line of best fit.

$$\text{slope} = \frac{\text{length of vertical leg}}{\text{length of horizontal leg}}$$

slope $= -\frac{2}{1}$, y-intercept $= 10$

equation for the line of best fit:

$y = -2x + 10$

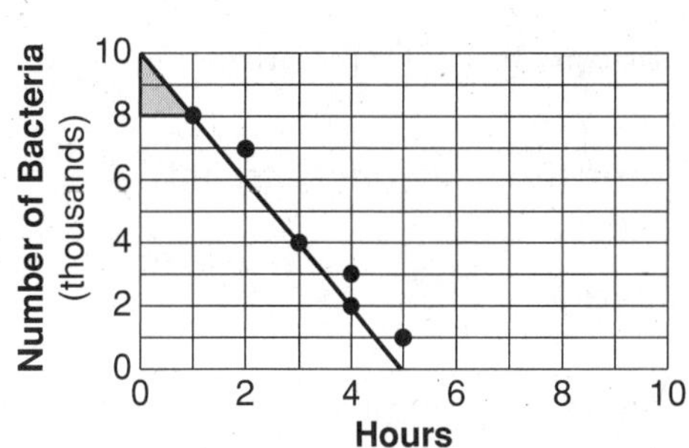

The data shows the number of bacteria present in a culture that has been treated with an anti-bacterial. The line of best fit can be used to predict the number of bacteria present after 2.5 hours.

$y = -2x + 10$

$y = -2(2.5) + 10$ Substitute $x = 2.5$.

$= -5 + 10$ Solve for y.

$= 5$

So, after 2.5 hours, the expected number of bacteria in the culture is approximately 5,000.

Find an equation for each line of best fit. Use the equation to answer the question.

3.

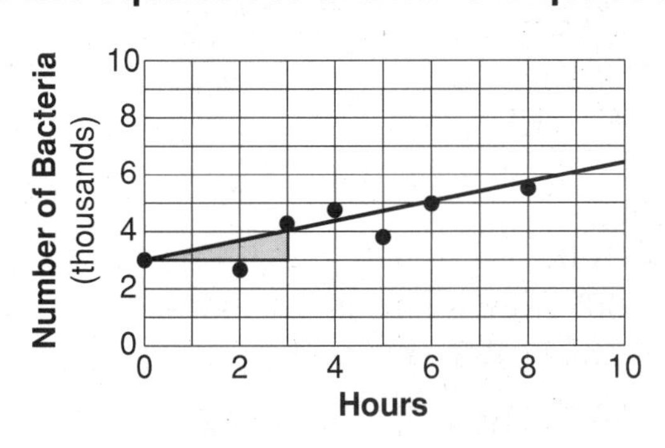

$m =$ ____, $b =$ ____,

equation: ______________

The expected number of bacteria after 10 hours is about __________.

4.

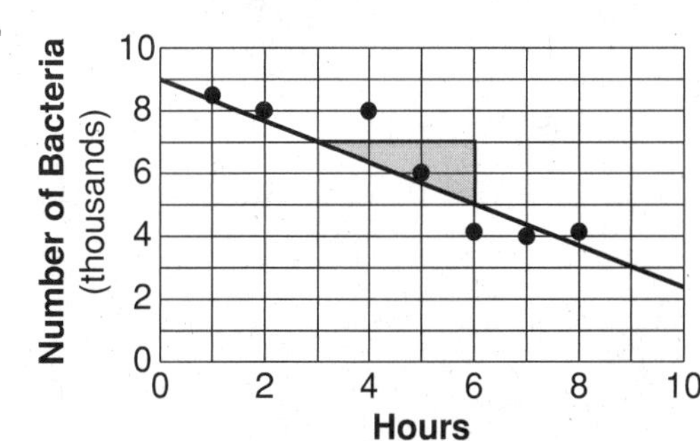

$m =$ ____, $b =$ ____,

equation: ______________

The expected number of bacteria after 10 hours is about __________.

Name ______________________ Date __________ Class __________

LESSON 12-7

Challenge

Use the Power of Technology

You can use a calculator to write an equation for the line of best fit. The following instructions are for the TI-83.

x	2	4	5	1	3	8	6	7
y	4	8	7	3	4	8	5	9

To enter the data in the calculator:

Display the Statistics menu. Press [STAT]

Choose the EDIT option. Press [ENTER]

Enter the *x*-values into list L1. Press 2 [ENTER] Press 4 [ENTER]

Move to list L2. Press [▶] Clear if necessary.

Enter the *y*-values into list L2. Press 4 [ENTER] Press 8 [ENTER]

To get the slope and *y*-intercept for the line of best fit for this data set:

Display the Y = editor. Press [Y =]

Display the Statistics menu. Press [STAT]

Select the 4th option from CALC. Press [▶] 4

Your screen reads **LinReg(*ax* + *b*).**

Attach L1 and L2 to your screen. Press [2nd] [L1] [,] [2nd] [L2] [,]

Attach Y1 to your screen. Press [VARS] [▶] [ENTER] 1

Get the slope and *y*-intercept. Press [ENTER]

The top part of your screen reads **LinReg**

y* = *ax* + *b

***a* = .7380952381** This is the slope.

***b* = 2.678571429** This is the *y*-intercept.

Use the information to write an equation for the line of best fit. $y = 0.74x + 2.68$

Compare these values to those in your text for Example 1.

Use a calculator to write an equation for the line of best fit for this data set.

x	3	8	4	4	5	7	1	6
y	5	14	8	9	12	3	2	11

____

Name ______________________________ Date ____________ Class ____________

LESSON 12-7

Problem Solving

Lines of Best Fit

Write the correct answer. Round to the nearest hundredth.

1. The table shows in what year different average speed barriers were broken at the Indianapolis 500. If x is the year, with $x = 0$ representing 1900, and y is the average speed, find the mean of the x- and y-coordinates.

   ____

Barrier (mi/h)	Year	Average Speed (mi/h)
80	1914	82.5
100	1925	101.1
120	1949	121.3
140	1962	140.9
160	1972	163.0
180	1990	186.0

2. Graph the data from exercise 1 and find the equation of the line of best fit.

   ____

   ____

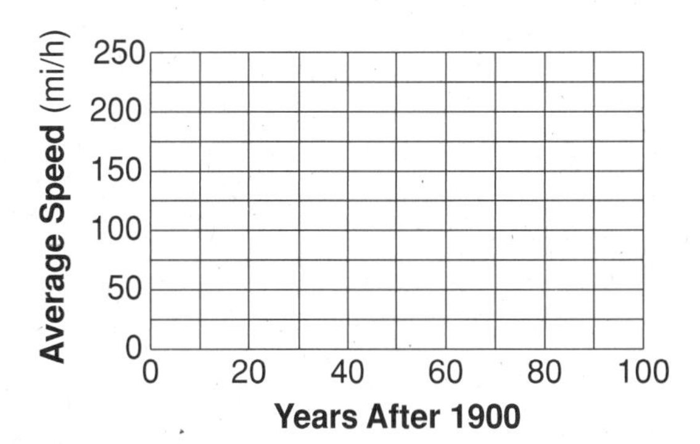

3. Use your equation to predict the year the 210 mph barrier will be broken.

   ____

The percent of the U.S. population who smokes can be represented by the line of best fit with the equation $y = -0.57x + 44.51$ where x is the year, $x = 0$ represents 1960, and y is the percent of the population who smokes. Circle the letter of the correct answer.

4. Which term describes the percent of the population that smokes?

   A Increasing **C** No change
   B Decreasing **D** Cannot tell

5. Use the equation to predict the percent of smokers in 2005.

   F 13.16% **H** 24.56%
   G 18.86% **J** 41.66%

6. Use the equation to predict when the percent of smokers will be less than 15%.

   A 1996 **C** 2012
   B 2010 **D** 2023

7. Use the equation to predict the percent of smokers in 2010.

   F 10.31% **H** 16.01%
   G 11.01% **J** 12.71%

Name ______________________ Date __________ Class __________

LESSON 12-7

Reading Strategies

Focus on Vocabulary

You may have T-shirts in different sizes, but none that really fits you. Your mom may tell you, “Choose the one that fits best.” She means the shirt that comes the closest to fitting you.

When the points on a scatter plot do not lie in a straight line, you can draw a **line of best fit.**

Draw a line of best fit so that there are about the same number of points above the line as below the line.

The points on this scatter plot show the number of sit-ups Tim completes each minute. The line of best fit shows **about** how many sit-ups Tim completes each minute.

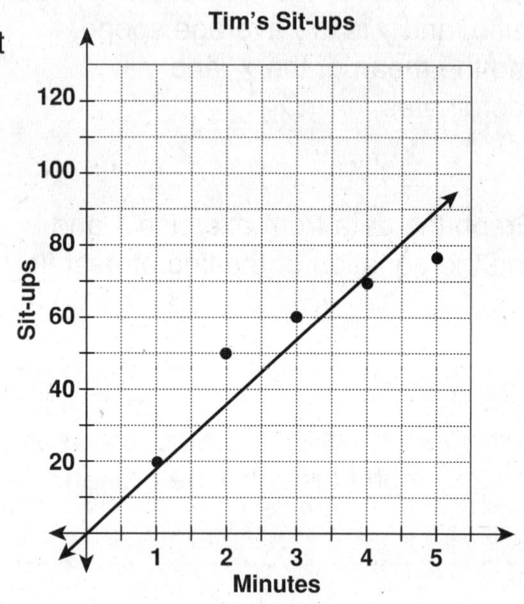

Use the scatter plot to answer the following questions.

1. What information is shown along the bottom edge of the scatter plot?

____

2. What information is shown along the left edge of the scatter plot?

____

3. Do the points on a scatter plot lie exactly on a straight line?

____

4. Does the line of best-fit show actual data?

____

5. Using the line of best fit, predict how many sit-ups Tim might complete in six minutes?

____

Name ______________________ Date __________ Class __________

LESSON 12-7

Puzzles, Twisters & Teasers

Ups and Downs!

Draw a line of best fit for each scatter plot. Select the answer that shows how many points fall above the line of best fit. To solve the riddle, match the letters of the answers with the blanks in the riddle.

1.

6	1	5	2
B	**D**	**J**	**L**

2.

2	1	5	3
L	**S**	**B**	**M**

3.

1	6	4	3
G	**W**	**K**	**F**

4.

3	4	5	1
H	**A**	**B**	**P**

5.

3	6	5	4
Y	**U**	**G**	**I**

6.

3	1	4	2
T	**V**	**N**	**K**

Where do fish like to go on vacation?

____ ____ ____ ____ ____ ____ ____
3 6 5 2 4 5 1

LESSON 12-1 **Practice A**
Graphing Linear Equations

Complete the tables for the given equations.

1. $y = 3x + 2$

x	3x + 2	y	(x, y)
−3	3(−3) + 2	−7	(−3, −7)
−2	3(−2) + 2	−4	(−2, −4)
−1	3(−1) + 2	−1	(−1, −1)
0	3(0) + 2	2	(0, 2)
1	3(1) + 2	5	(1, 5)
2	3(2) + 2	8	(2, 8)
3	3(3) + 2	11	(3, 11)

2. $y = -x - 3$

x	−x − 3	y	(x, y)
−3	−(−3) − 3	0	(−3, 0)
−2	−(−2) − 3	−1	(−2, −1)
−1	−(−1) − 3	−2	(−1, −2)
0	−(0) − 3	−3	(0, −3)
1	−(1) − 3	−4	(1, −4)
2	−(2) − 3	−5	(2, −5)
3	−(3) − 3	−6	(3, −6)

Complete the table for the given equation. Then graph the equation.

3. $2x - y = 1$

x	2x − 1	y	(x, y)
−3	2(−3) − 1	−7	(−3, −7)
−2	2(−2) − 1	−5	(−2, −5)
−1	2(−1) − 1	−3	(−1, −3)
0	2(0) − 1	−1	(0, −1)
1	2(1) − 1	1	(1, 1)
2	2(2) − 1	3	(2, 3)
3	2(3) − 1	5	(3, 5)

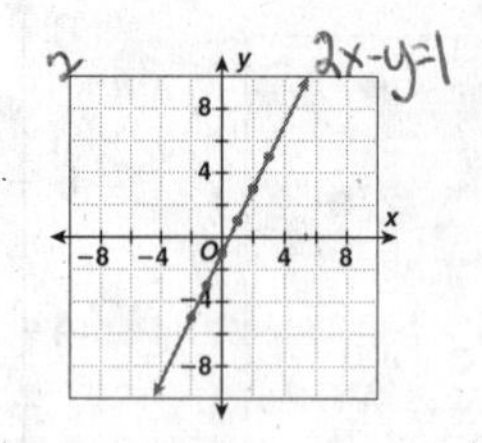

4. Is the equation in Exercise 3 linear? Explain. **Possible answer:**
Yes, the points form a straight line and each time x increases by 1 unit, y increases by 2 units.

LESSON 12-1 **Practice B**
Graphing Linear Equations

Graph each equation and tell whether it is linear.

1. $y = -2x - 5$

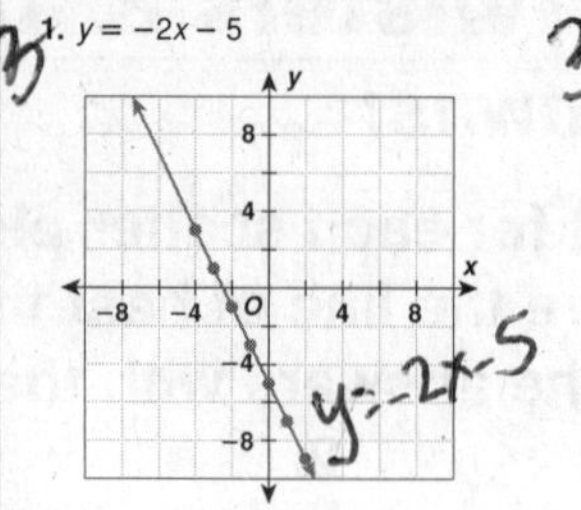

linear

2. $y = -x^2 + 1$

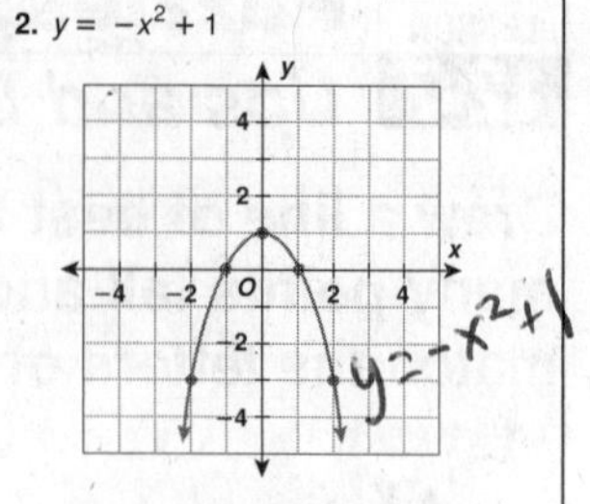

not linear

3. $y = x^2 - 7$

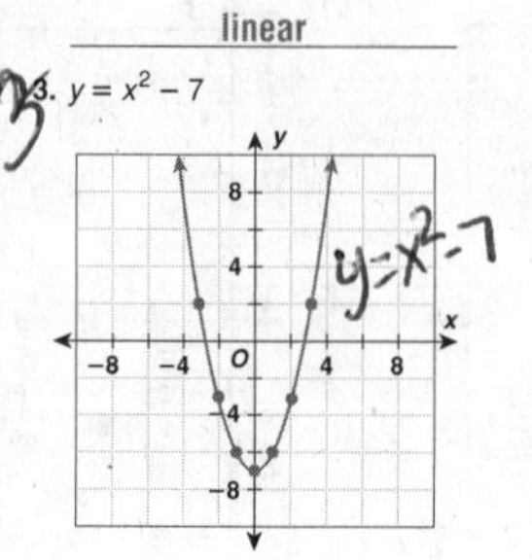

not linear

4. $y = \frac{1}{2}x - 1$

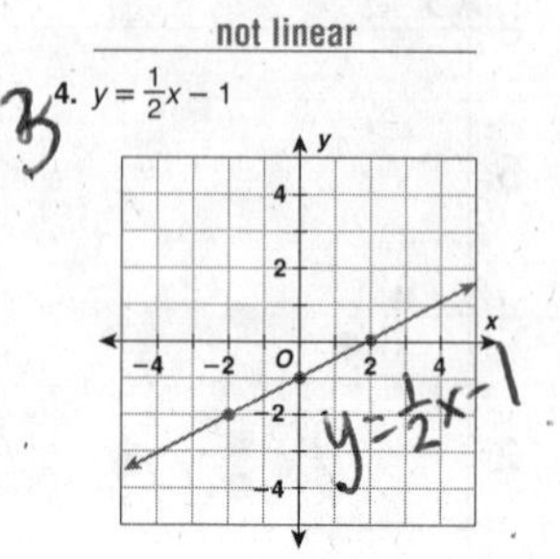

linear

5. A real estate agent commission may be based on the equation $C = 0.06s + 450$, where s represents the total sales. If the agent sells a property for $125,000, what is the commission earned by the agent? Graph the equation and tell whether it is linear.

$7950; linear

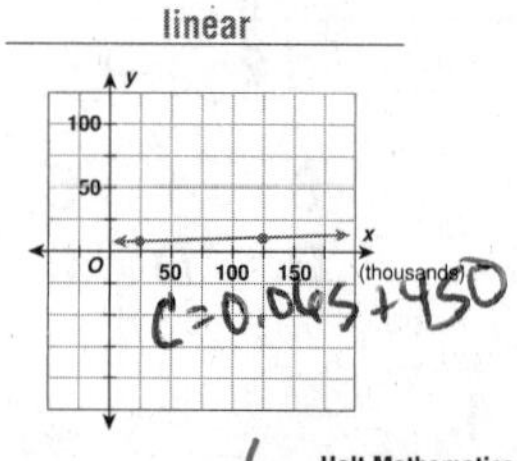

LESSON 12-1 **Practice C**
Graphing Linear Equations

Graph each equation and tell whether it is linear.

1. $y = -3x$

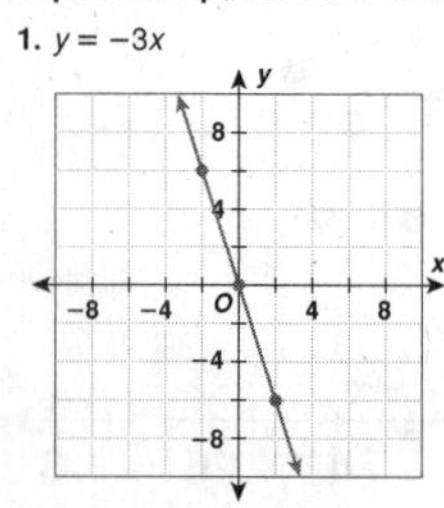

2. $y = x - 5$

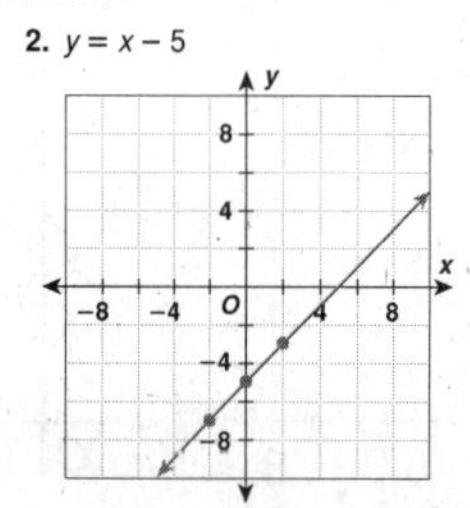

3. $y = \frac{1}{2}x - 3$

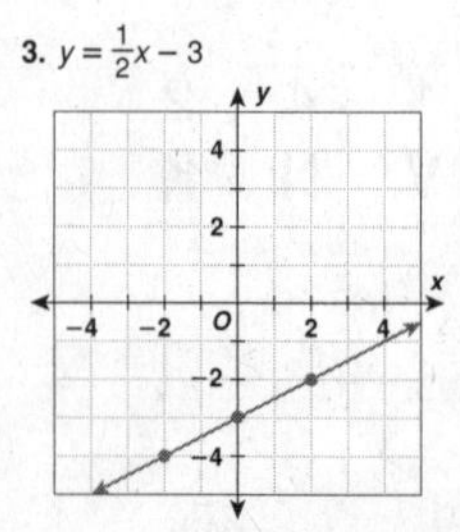

4. $y = -x^2 - 2$

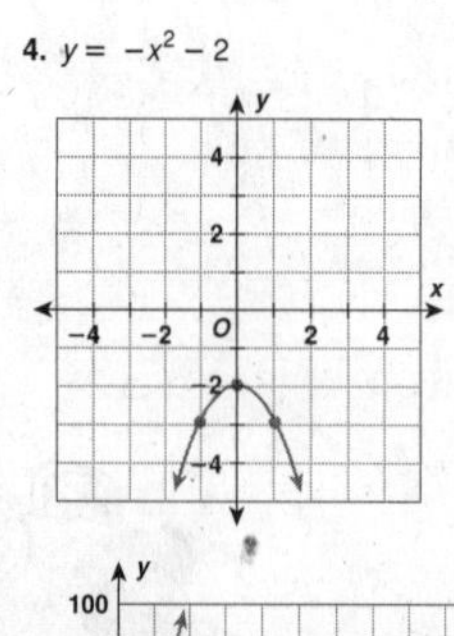

5. Mrs. Blanche grades a math test with a curve based on the formula $G = 5.5p + 10$, where G is the curved grade and p represents the number of problems correct. If Sebastian had 10 problems correct, Alisha had 12 problems correct, and Miguel had 14 problems correct, what was each student's curved grade? Graph the equation and tell whether it is linear.

Sebastian: 65; Alisha: 76; Miguel: 87; linear

LESSON 12-1 **Reteach**
Graphing Linear Equations

The graph of a **linear equation** is a straight line.

The line shown is the graph of $y = \frac{3}{2}x + 1$.

All the points on the line are solutions of the equation.

Each time the x-value increases by 2, the y-value increases by 3. So, a constant change in the x-value corresponds to a constant change in the y-values.

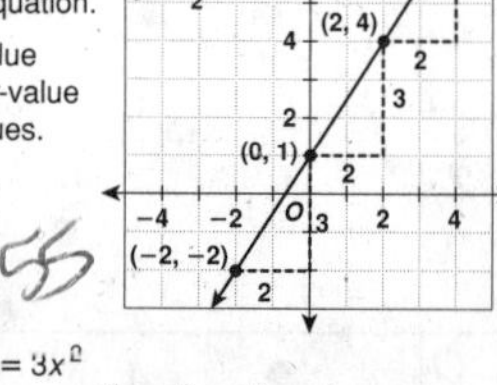

$y = 3x - 4$

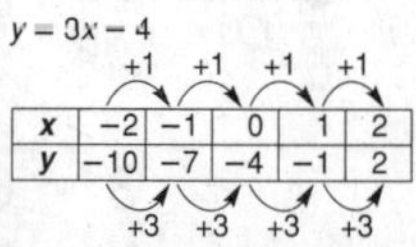

x	−2	−1	0	1	2
y	−10	−7	−4	−1	2

+1 +1 +1 +1 (x); +3 +3 +3 +3 (y)

Since a constant change in the x-value corresponds to a constant change in the y-value, $y = 3x - 4$ is a linear equation.

$y = 3x^2$

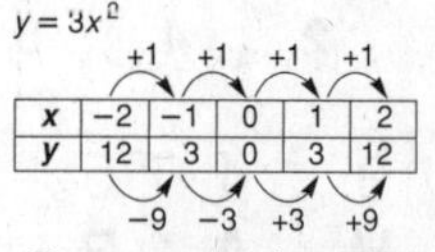

x	−2	−1	0	1	2
y	12	3	0	3	12

+1 +1 +1 +1 (x); −9 −3 +3 +9 (y)

Since a constant change in the x-value does not correspond to a constant change in the y-value, $y = 3x^2$ is not a linear equation.

Each equation has a table of solutions. Indicate the changes in x-values and in y-values. Tell whether the equation is linear.

1. $y = 2x - 5$

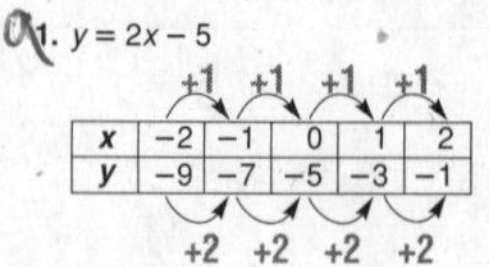

x	−2	−1	0	1	2
y	−9	−7	−5	−3	−1

+1 +1 +1 +1; +2 +2 +2 +2

linear

2. $y = 2x^3$

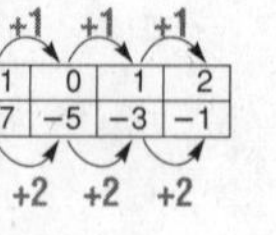

x	−2	−1	0	1	2
y	−16	−2	0	2	16

+1 +1 +1 +1; +14 +2 +2 +14

not linear

LESSON 12-1 **Reteach**

Graphing Linear Equations (continued)

To graph a linear equation, make a table to find several solutions. Choose *x*-values that are easy to graph. Substitute each *x*-value into the equation to find the corresponding *y*-value. Plot your solutions and draw a line connecting them.

x	$\frac{1}{2}x - 3$	y	(x, y)
−4	$\frac{1}{2}(-4) - 3$	−5	(−4, −5)
−2	$\frac{1}{2}(-2) - 3$	−4	(−2, −4)
0	$\frac{1}{2}(0) - 3$	−3	(0, −3)
6	$\frac{1}{2}(6) - 3$	0	(6, 0)

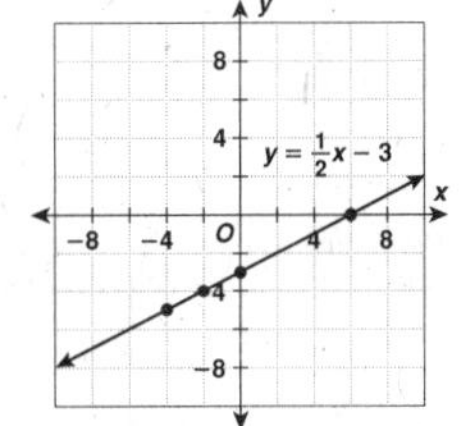

Complete the table for each equation and then graph the equation.

3. $y = 3x - 1$

x	$3x - 1$	y	(x, y)
−2	3(−2) − 1	−7	(−2, −7)
0	3(0) − 1	−1	(0, −1)
1	3(1) − 1	2	(1, 2)
3	3(3) − 1	8	(3, 8)

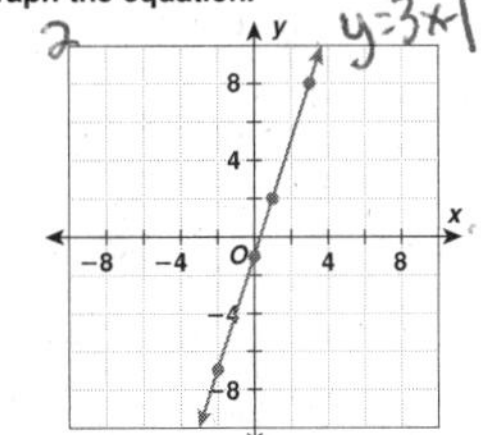

4. $y = -\frac{3}{2}x + 1$

x	$-\frac{3}{2}x + 1$	y	(x, y)
−2	$-\frac{3}{2}(-2) + 1$	4	(−2, 4)
0	$-\frac{3}{2}(0) + 1$	1	(0, 1)
2	$-\frac{3}{2}(2) + 1$	−2	(2, −2)
4	$-\frac{3}{2}(4) + 1$	−5	(4, −5)

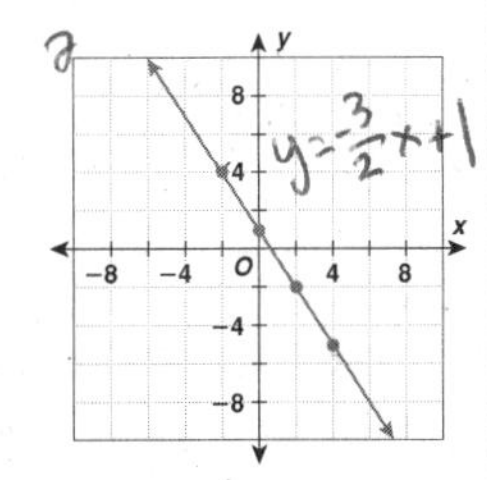

LESSON 12-1 **Challenge**

A Recognition Factor

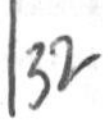

Different kinds of equations have different kinds of graphs. By studying the graphs of different kinds of equations, you can learn to recognize characteristics of the equations.

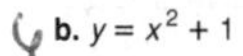

1. Complete the table of values to graph each equation. Draw all the graphs on the given grid. Write each equation near its graph.

a. $y = 2x + 1$

x	y
−4	−7
−3	−5
−2	−3
−1	−1

b. $y = x^2 + 1$

x	y
−1	2
0	1
1	2
2	5

c. $xy = 6$

x	y
1	6
2	3
3	2
6	1

d. $x + y = -1$

x	y
−5	4
−4	3
−3	2
−2	1

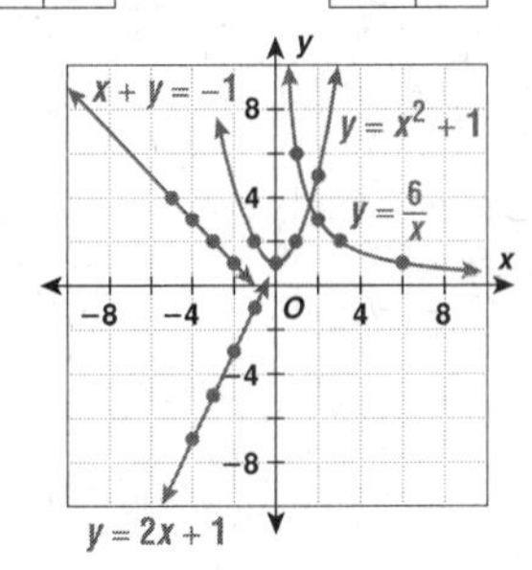

2. Analyze the equations in Exercise 1 and your graphs of the equations. Make a conjecture about how you might recognize a linear equation without graphing it.

Possible answer: In a linear equation, both variables are raised to the first power. Also the variable terms are separated by + or −, not × or ÷ .

LESSON 12-1 **Problem Solving**

Graphing Linear Equations

Write the correct answer.

1. The distance in feet traveled by a falling object is found by the formula $d = 16t^2$ where *d* is the distance in feet and *t* is the time in seconds. Graph the equation. Is the equation linear?

The equation is not linear.

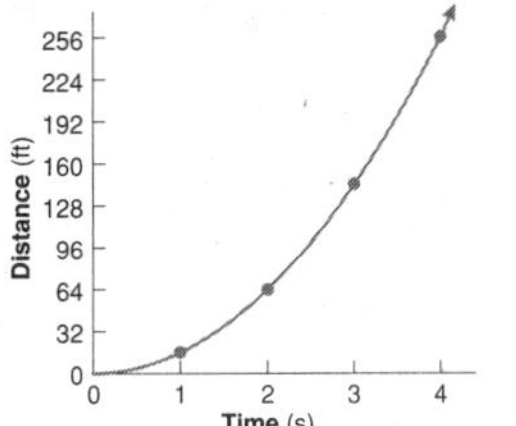

2. The formula that relates Celsius to Fahrenheit is $F = \frac{9}{5}C + 32$. Graph the equation. Is the equation linear?

The equation is linear.

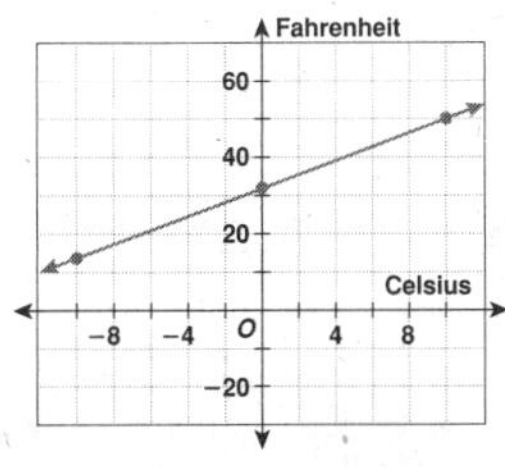

Wind chill is the temperature that the air feels like with the effect of the wind. The graph below shows the wind chill equation for a wind speed of 25 mph. For Exercises 3–6, refer to the graph.

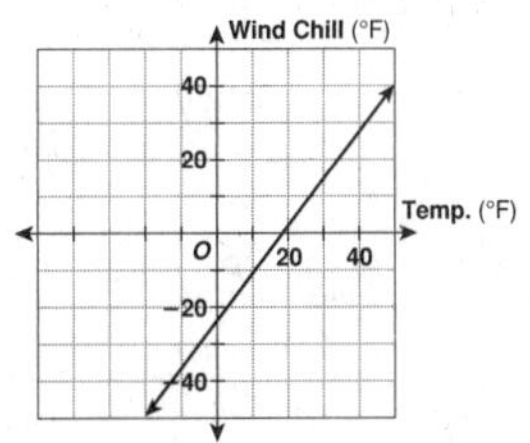

3. If the temperature is 40° with a 25 mph wind, what is the wind chill?
   A 6° (C) 29°
   B 20° D 40°

4. If the temperature is 20° with a 25 mph wind, what is the wind chill?
   (F) 3° H 13°
   G 10° J 20°

5. If the temperature is 0° with a 25 mph wind, what is the wind chill?
   A −30° C −15°
   (B) −24° D 0°

6. If the wind chill is 10° and there is a 25 mph wind, what is the actual temperature?
   F −11° H 15°
   G 0° (J) 25°

LESSON 12-1 **Reading Strategies**

Use a Graphic Organizer

This graphic organizer can help you understand linear equations.

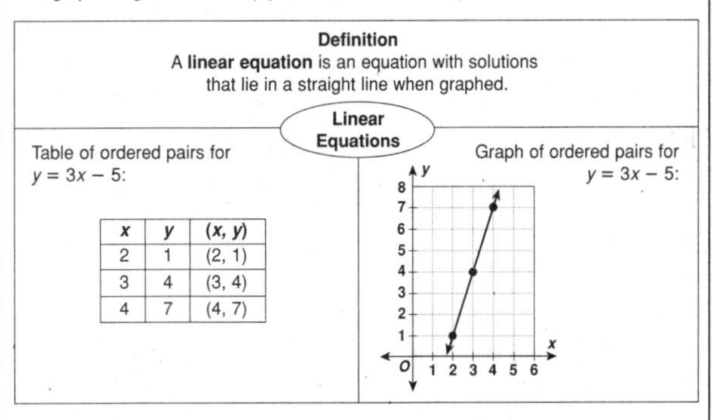

Definition
A **linear equation** is an equation with solutions that lie in a straight line when graphed.

Linear Equations

Table of ordered pairs for $y = 3x - 5$:

x	y	(x, y)
2	1	(2, 1)
3	4	(3, 4)
4	7	(4, 7)

Graph of ordered pairs for $y = 3x - 5$:

Use the information in the graphic organizer to answer the following questions.

1. What is a linear equation?
   an equation with solutions that lie in a straight line
2. Write the linear equation shown in the example above.
   $y = 3x - 5$
3. How many solutions for the equation are plotted on the graph?
   3
4. How is each ordered pair shown on the graph?
   with a point
5. Write one of the solutions (ordered pairs) shown on the graph.
   (2, 1), (3, 4), or (4, 7)
6. What figure is formed when the points on the graph are connected?
   a straight line

LESSON 12-1

Puzzles, Twisters & Teasers

Straight and Narrow?

Decide whether each graph is linear or nonlinear. Circle the letter above your answer. Use the letters to solve the riddle.

1. 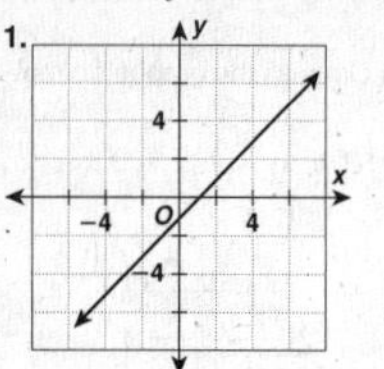

I linear | Q nonlinear

2. 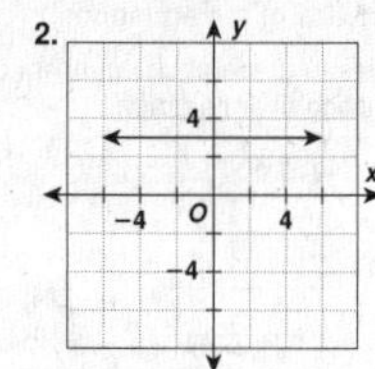

A linear | Z nonlinear

3. 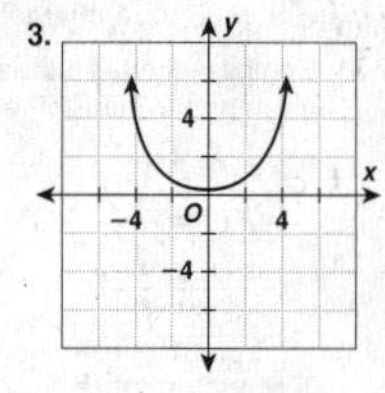

X linear | N nonlinear

4. 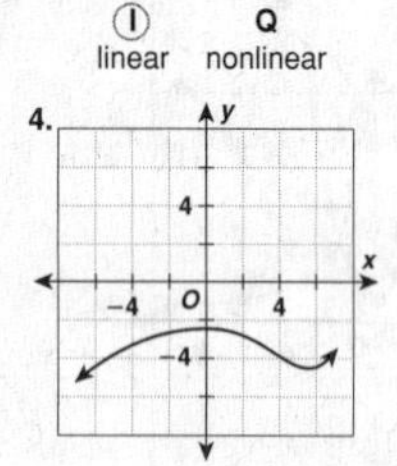

Y linear | L nonlinear

5. 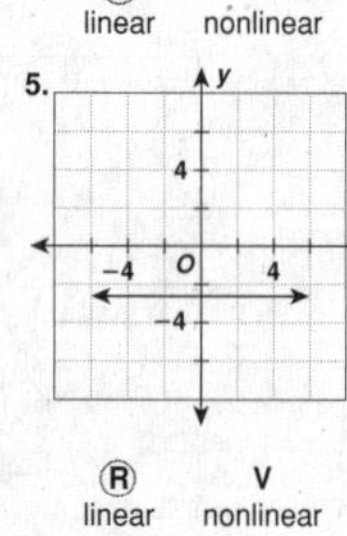

R linear | V nonlinear

6. 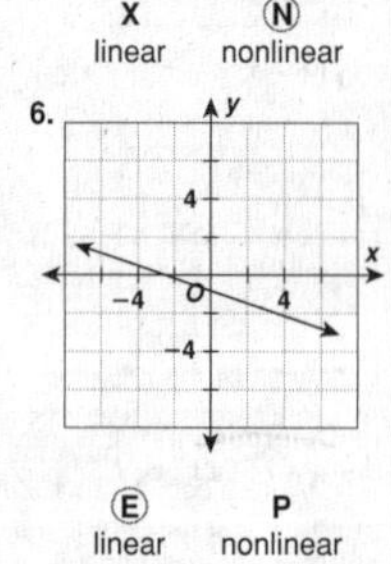

E linear | P nonlinear

What do you call a dancing sheep?

A B $\underset{2}{\underline{A}}$ $\underset{2}{\underline{A}}$ – $\underset{4}{\underline{L}}$ $\underset{4}{\underline{L}}$ $\underset{6}{\underline{E}}$ $\underset{5}{\underline{R}}$ $\underset{1}{\underline{I}}$ $\underset{3}{\underline{N}}$ $\underset{2}{\underline{A}}$

LESSON 12-2

Practice A

Slope of a Line

Find the slope of the line that passes through each pair of points.

1. (1, 0), (2, 4) — 4
2. (6, 2), (2, −2) — 1
3. (−1, 1), (4, 4) — $\frac{3}{5}$
4. (−7, 4), (2, 1) — $-\frac{1}{3}$
5. (5, −3), (−2, −3) — 0
6. (−3, 2), (2, 7) — 1

Determine whether each graph shows a constant or variable rate of change. Explain your reasoning.

7. 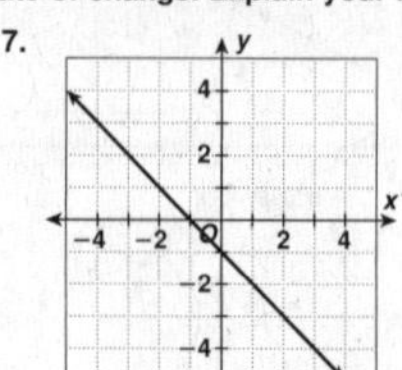

constant; The slope between any two points is always the same.

8. 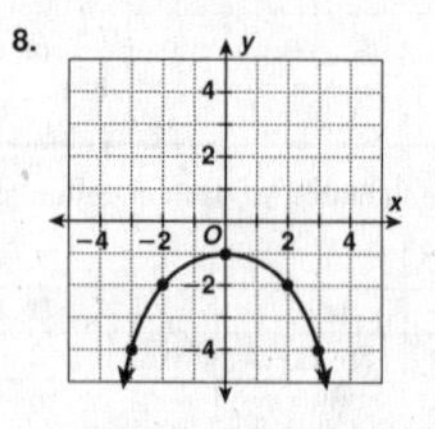

variable; The slope varies from point to point.

9. 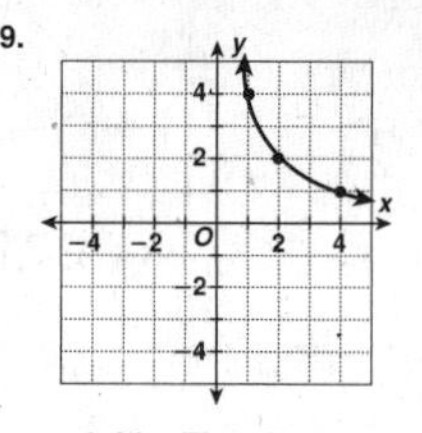

variable; The slope varies from point to point.

10. constant; The slope between any two points is always the same.

LESSON 12-2

Practice B

Slope of a Line

Find the slope of the line that passes through each pair of points.

1. (−2, −8), (1, 4) — 4
2. (−2, 0), (0, 4), — 2
3. (0, 4), (4, 4) — 0
4. (3, −6), (2, −4) — −2
5. (−3, 4), (3, −4) — $-\frac{4}{3}$
6. (3, 0), (0, −6), — 2
7. (3, 2), (3, −2) — undefined
8. (−4, 4), (3, −1) — $-\frac{5}{7}$

Determine whether each graph shows a constant or variable rate of change. Explain your reasoning.

9. 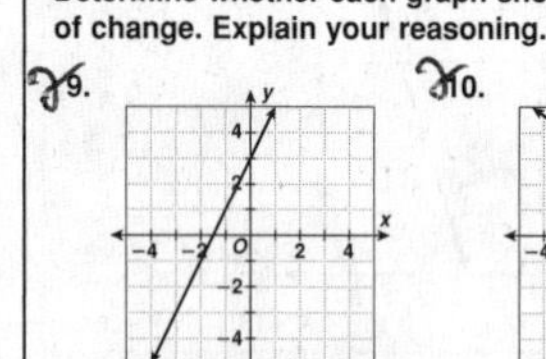

constant; The slope between any two points is always the same.

10. 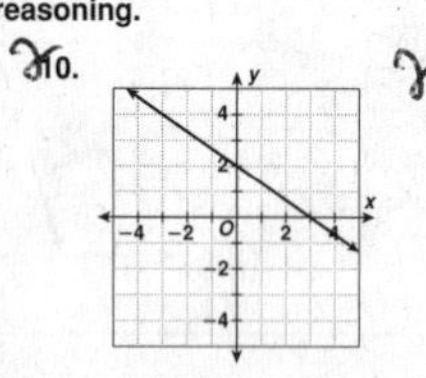

constant; The slope between any two points is always the same.

11. 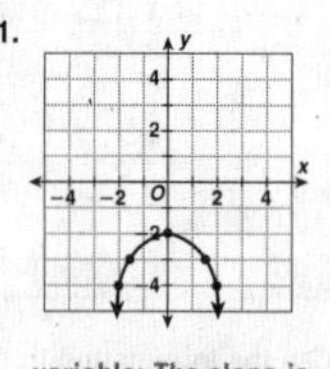

variable; The slope is positive in Quadrant III and negative in Quadrant IV.

12. The table shows the distance Ms. Long had traveled as she went to the beach. Use the data to make a graph. Find the slope of the line and explain what it shows.

Time (min)	Distance (mi)
8	6
12	9
16	12
20	15

The slope is $\frac{3}{4}$, which means that for every 4 minutes Ms. Long drives, she travels 3 miles. She is driving 45 mph.

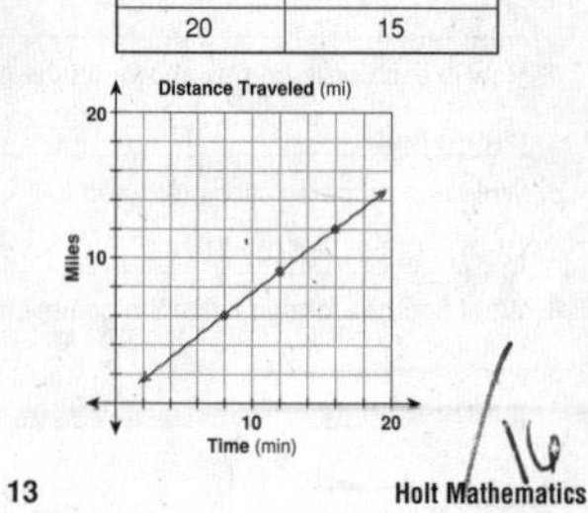

LESSON 12-2

Practice C

Slope of a Line

For exercises 1–4, use the graph to find the slope of the line.

1. 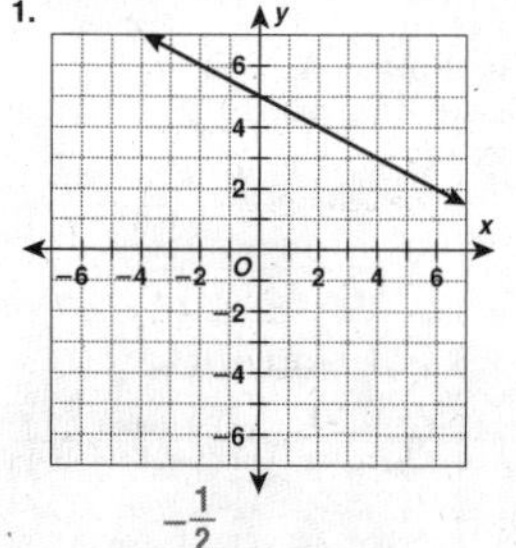

$-\frac{1}{2}$

2. 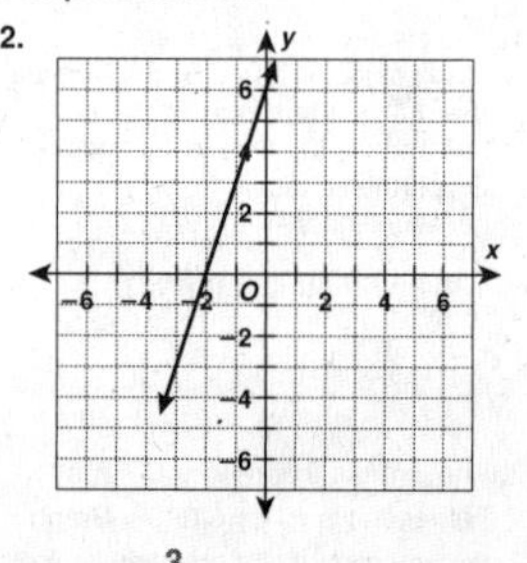

3

3. 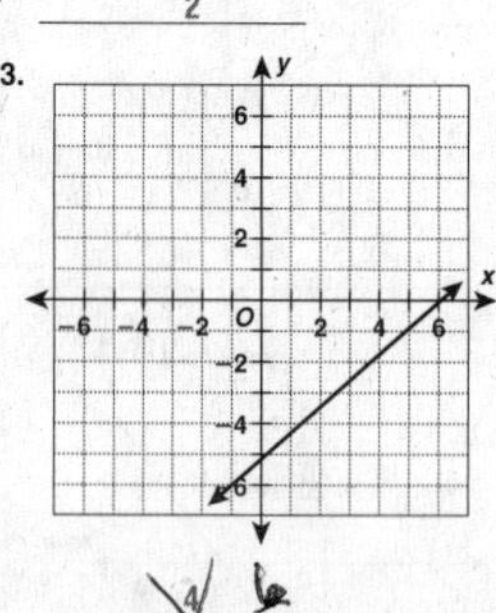

$\frac{4}{5}$

4. 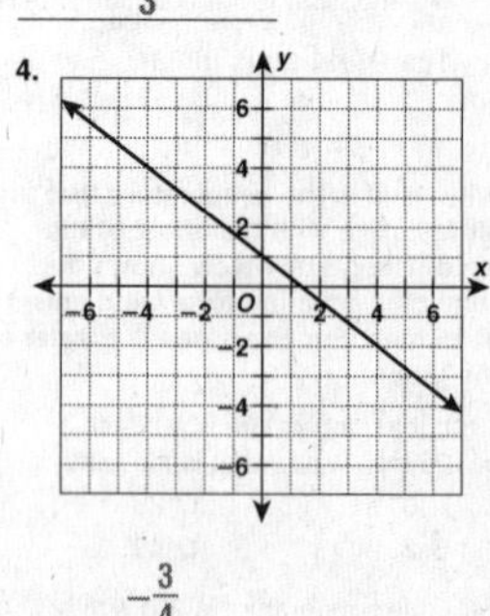

$-\frac{3}{4}$

5. If a quadrilateral has vertices $A(-2, 2)$, $B(2, 3)$, $C(3, -4)$, and $D(-3, -2)$, find the slope of $\overline{AB}$, $\overline{BC}$, $\overline{CD}$, and $\overline{DA}$.

slope $\overline{AB}$: $\frac{1}{4}$; slope $\overline{BC}$: −7; slope $\overline{CD}$: $-\frac{1}{3}$; slope $\overline{DA}$: 4

LESSON **12-2**

Reteach
Slope of a Line

The **slope** of a line is a measure of its tilt, or slant.

The slope of a straight line is a constant ratio, the "rise over run," or the **vertical change** over the **horizontal change**.

You can find the slope of a line by comparing any two of its points. The vertical change is the difference between the two *y*-values. And the horizontal change is the difference between the two *x*-values.

slope $= \frac{y_2 - y_1}{x_2 - x_1}$

point *A*: (3, 2) point *B*: (4, 4)

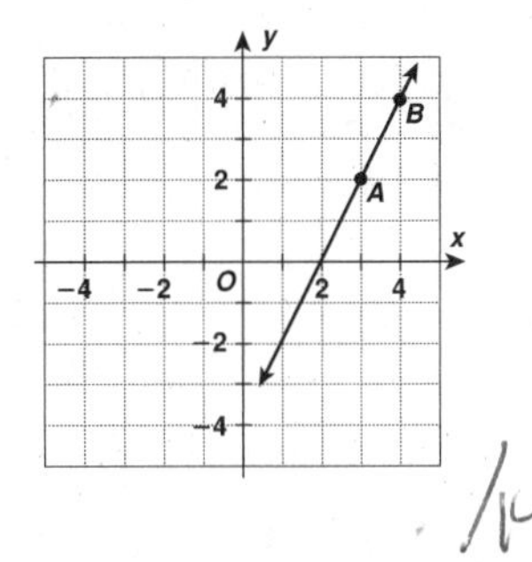

Make point *A* (x_1, y_1).
Make point *B* (x_2, y_2).

slope $= \frac{4 - 2}{4 - 3}$

$= \frac{2}{1}$, or 2

So, the slope of the line is 2.

You can make point *A* (x_2, y_2) and point *B* (x_1, y_1).

slope $= \frac{2 - 4}{3 - 4}$

$= \frac{-2}{-1}$, or 2

So, the slope remains 2.

Find the slope of the line that passes through each pair of points.

1. (1, 5) and (2, 6) ___1___
2. (0, 3) and (2, 7) ___2___
3. (2, 5) and (3, 4) ___−1___
4. (6, 9) and (2, 7) ___$\frac{1}{2}$___
5. (6, 5) and (8, −1) ___−3___
6. (7, −4) and (4, −2) ___$-\frac{2}{3}$___

LESSON **12-2**

Reteach
Slope of a Line (continued)

A straight line has a constant slope, so it shows a **constant rate of change**. The same change in *y* always results in the same change in *x*.

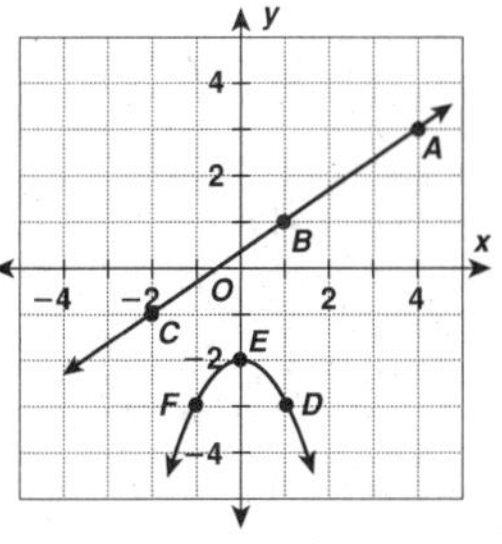

From point *C* to point *B*:

$\frac{\text{change in } y}{\text{change in } x} = \frac{2}{3}$

From point *B* to point *A*:

$\frac{\text{change in } y}{\text{change in } x} = \frac{2}{3}$

A curved line doesn't have a constant slope, so it shows a **variable rate of change**.

Between point *F* and point *E*, the curved line has a positive slope.
Between point *E* and point *D*, the curved line has a negative slope.

So, the curved line has a variable rate of change.

Determine whether each graph shows a constant or a variable rate of change. Write *constant* or *variable*.

7.

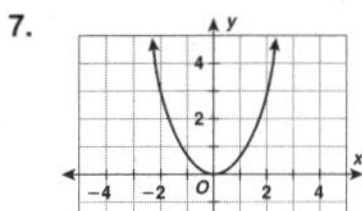

___variable___

8.

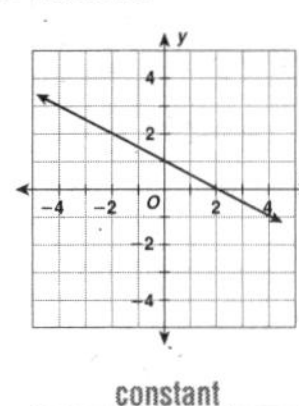

___constant___

9.

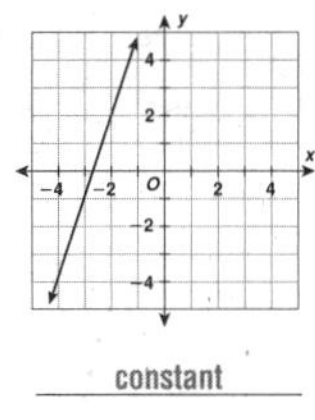

___constant___

LESSON **12-2**

Challenge
Aligned?

1. Points *A*, *B*, and *C* are on the same line. Draw a conclusion about the slope between *A* and *B* and the slope between *B* and *C*.

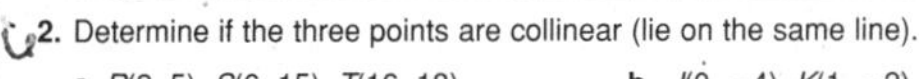

slope between *A* and *B* = slope between *B* and *C*

2. Determine if the three points are collinear (lie on the same line).

a. *R*(2, 5), *S*(6, 15), *T*(16, 18)

slope between *R* and *S* =

$\frac{15 - 5}{6 - 2} = \frac{10}{4} = \frac{5}{2}$

slope between *S* and *T* =

$\frac{18 - 15}{16 - 6} = \frac{3}{10}$

R, *S*, *T* ___are not___ collinear.

b. *J*(0, −4), *K*(1, −2), *L*(3, 2)

slope between *J* and *K* =

$\frac{-2 - (-4)}{1 - 0} = \frac{-2 + 4}{1} = 2$

slope between *K* and *L* =

$\frac{2 - (-2)}{3 - 1} = \frac{2 + 2}{2} = \frac{4}{2} = 2$

J, *K*, *L* ___are___ collinear.

3. Find the value of *k* so that *U*(−5, −1), *V*(−1, −5), and *W*(5, *k*) are collinear.

a. Find the slope between *U* and *V*. $\frac{-5 - (-1)}{-1 - (-5)} = \frac{-5 + 1}{-1 + 5} = \frac{-4}{4} = -1$

b. Find the slope between *V* and *W*. $\frac{k - (-5)}{5 - (-1)} = \frac{k + 5}{5 + 1} = \frac{k + 5}{6}$

c. Set the results of parts a and b equal to each other and solve for *k*. Justify your result.

$\frac{-1}{1} = \frac{k + 5}{6}$

$(k + 5)(1) = (-1)(6)$

$k + 5 = -6$

$k = -11$

Check: When $k = -11$, the slope between *V* and *W* should equal −1.

$\frac{k + 5}{6} = \frac{-11 + 5}{6} = \frac{-6}{6} = -1$

4. The points *P*(2, −3), *Q*(2, 3) and *R*(*k*, 0) are collinear. Find *k*. Justify your result.

Since *P* and *Q* have the same *x*-values, $\overleftrightarrow{PQR}$ is a vertical line. So, $k = 2$.

LESSON **12-2**

Problem Solving
Slope of a Line

Write the correct answer.

1. The state of Kansas has a fairly steady slope from the east to the west. At the eastern side, the elevation is 771 ft. At the western edge, 413 miles across the state, the elevation is 4039 ft. What is the approximate slope of Kansas?

___−0.0015___

2. The Feathered Serpent Pyramid in Teotihuacan, Mexico, has a square base. From the center of the base to the center of an edge of the pyramid is 32.5 m. The pyramid is 19.4 m high. What is the slope of each face of the pyramid?

___$\frac{19.4}{32.5}$___

3. On a highway, a 6% grade means a slope of 0.06. If a highway covers a horizontal distance of 0.5 miles and the elevation change is 184.8 feet, what is the grade of the road? (Hint: 5280 feet = 1 mile.)

___7%___

4. The roof of a house rises vertically 3 feet for every 12 feet of horizontal distance. What is the slope, or pitch of the roof?

___$\frac{1}{4}$___

Use the graph for Exercises 5–8.

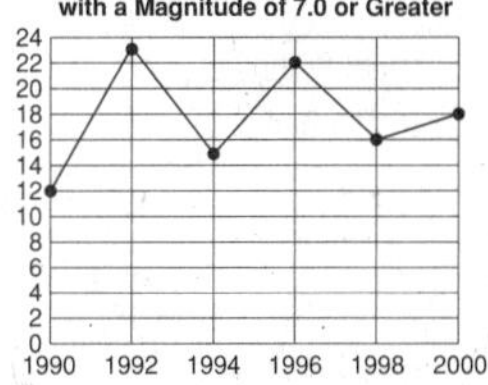

5. Find the slope of the line between 1990 and 1992.

A $\frac{2}{11}$ **(C)** $\frac{11}{2}$
B $\frac{35}{3982}$ D $\frac{11}{1992}$

6. Find the slope of the line between 1994 and 1996.

(F) $\frac{7}{2}$ H $\frac{2}{7}$
G $\frac{37}{3990}$ J $\frac{7}{1996}$

7. Find the slope of the line between 1998 and 2000.

(A) 1
B $\frac{1}{999}$
C $\frac{1}{1000}$
D 2

8. What does it mean when the slope is negative?

F The number of earthquakes stayed the same.
G The number of earthquakes increased.
(H) The number of earthquakes decreased.
J It means nothing.

LESSON 12-2 **Reading Strategies**

Use a Graphic Organizer

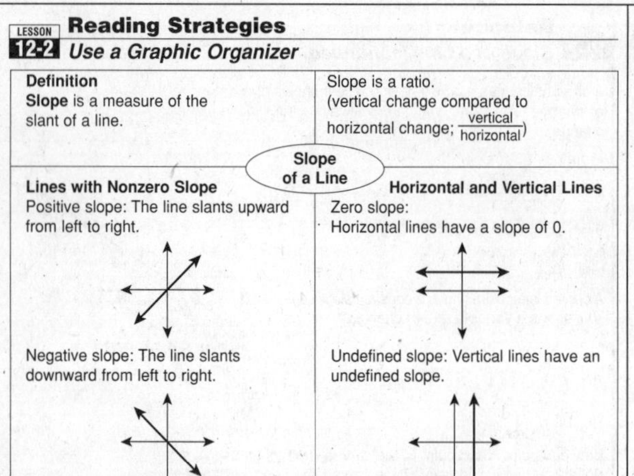

Definition **Slope** is a measure of the slant of a line.	Slope is a ratio. (vertical change compared to horizontal change; $\frac{\text{vertical}}{\text{horizontal}}$)
Lines with Nonzero Slope Positive slope: The line slants upward from left to right. Negative slope: The line slants downward from left to right.	**Horizontal and Vertical Lines** Zero slope: Horizontal lines have a slope of 0. Undefined slope: Vertical lines have an undefined slope.

Use the graphic organizer to answer the following questions.

1. What do you call the slant of a line?

   slope

2. Write the ratio that is used to describe slope.

   $\frac{\text{vertical change}}{\text{horizontal change}}$

3. How can you tell if a line has positive slope?

   The line slants upward from left to right.

4. How you can tell if a line has negative slope?

   The line slants downward from left to right.

5. What kind of line has a slope of 0?

   a horizontal line

LESSON 12-2 **Puzzles, Twisters & Teasers**

A Slippery Slope!

Determine which kind of slope each line has. Then use the letters of the correct answers to solve the riddle.

1.

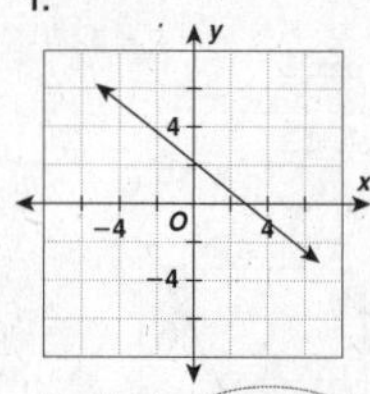

Positive **B** Negative **O** Zero **D** Undefined **L**

2.

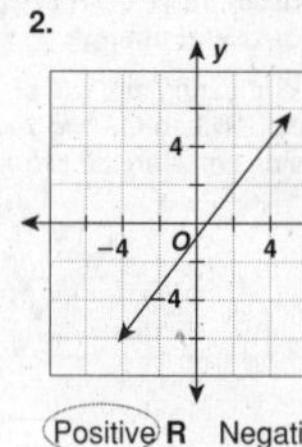

Positive **R** Negative **C** Zero **Q** Undefined **D**

3.

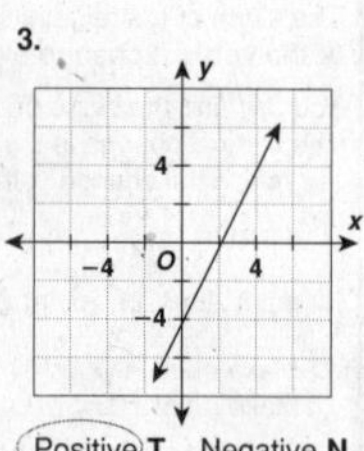

Positive **T** Negative **N** Zero **G** Undefined **N**

4.

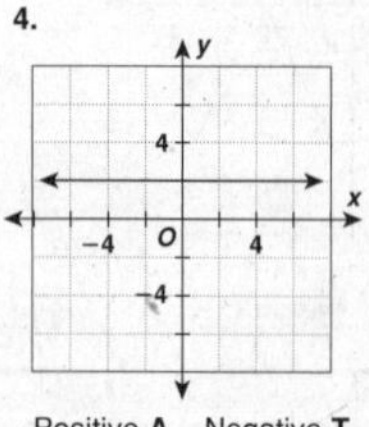

Positive **A** Negative **T** Zero **H** Undefined **V**

5.

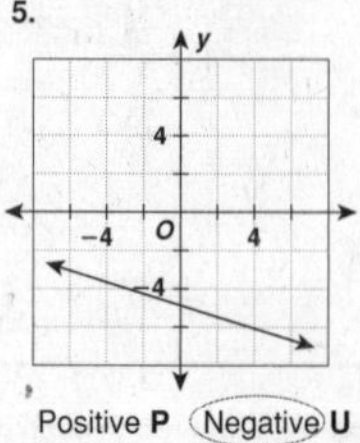

Positive **P** Negative **U** Zero **F** Undefined **Q**

6.

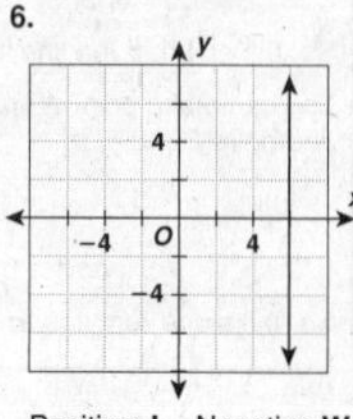

Positive **J** Negative **W** Zero **X** Undefined **Y**

What time do you go to the dentist?

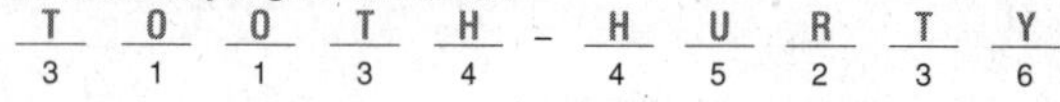

T	O	O	T	H	–	H	U	R	T	Y
3	1	1	3	4		4	5	2	3	6

LESSON 11-3 **Practice A**

Using Slopes and Intercepts

1. Name the ordered pair if the x-intercept is −2. **(−2, 0)**
2. Name the ordered pair if the y-intercept is 8. **(0, 8)**
3. In the ordered pair (9, 0), what is the x-intercept? **9**
4. In the ordered pair (0, 0), what is the relationship of the x-intercept and y-intercept? **x-intercept = y-intercept**

Find the x-intercept and y-intercept of each line. Use the intercepts to graph the equation.

5. $x + y = 5$

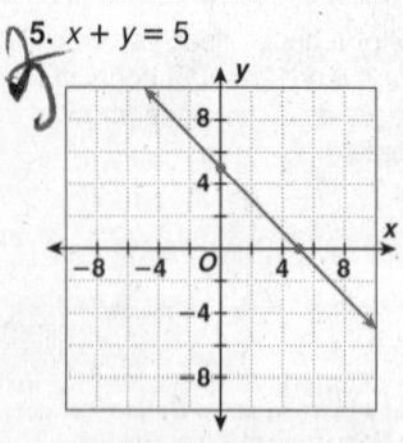

x-intercept is 5; y-intercept is 5

6. $2x - y = 6$

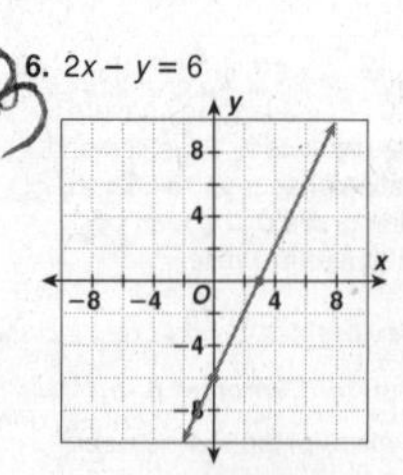

x-intercept is 3; y-intercept is −6

Write each equation in slope-intercept form, and then find the slope and y-intercept.

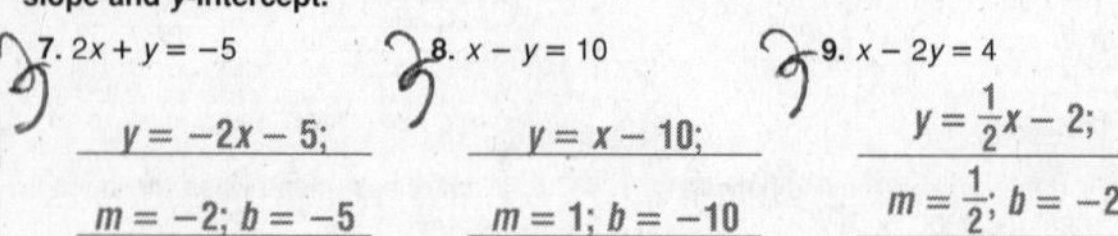

7. $2x + y = -5$

   $y = -2x - 5$; $m = -2$; $b = -5$

8. $x - y = 10$

   $y = x - 10$; $m = 1$; $b = -10$

9. $x - 2y = 4$

   $y = \frac{1}{2}x - 2$; $m = \frac{1}{2}$; $b = -2$

Write the equation of the line that passes through each pair of points in slope-intercept form.

10. (1, 2), (−1, 0) $y = x + 1$
11. (1, −3), (−1, 1) $y = -2x - 1$
12. (1, 1), (−3, −3) $y = x$

LESSON 12-3 **Practice B**

Using Slopes and Intercepts

Find the x-intercept and y-intercept of each line. Use the intercepts to graph the equation.

1. $x - y = -3$

   x-intercept is −3; y-intercept is 3

2. $2x + 3y = 12$

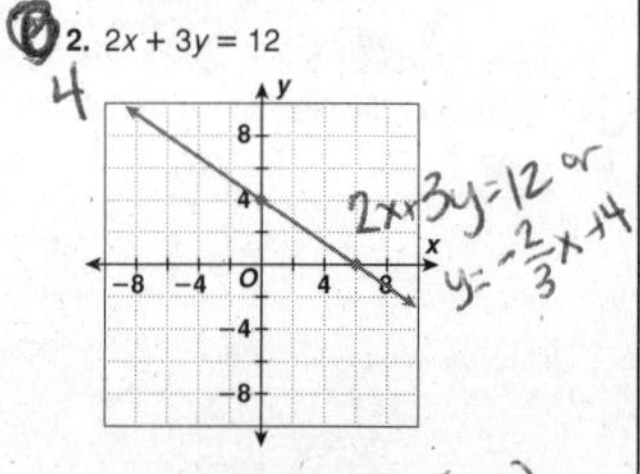

   x-intercept is 6; y-intercept is 4

Write each equation in slope-intercept form, and then find the slope and y-intercept.

3. $3x + y = 0$

   $y = -3x$; $m = -3$; $b = 0$

4. $2x - y = -15$

   $y = 2x + 15$; $m = 2$; $b = 15$

5. $x - 5y = 10$

   $y = \frac{1}{5}x - 2$; $m = \frac{1}{5}$; $b = -2$

Write the equation of the line that passes through each pair of points in slope-intercept form.

6. (3, 4), (4, 6) $y = 2x - 2$
7. (−1, −1), (2, −10) $y = -3x - 4$
8. (6, 5), (−9, −20) $y = \frac{5}{3}x - 5$

9. A pizzeria charges $8 for a large cheese pizza, plus $2 for each topping. The total cost for a large pizza is given by the equation C = 2t + 8, where t is the number of toppings. Identify the slope and y-intercept, and use them to graph the equation for t between 0 and 5 toppings.

   The slope is 2, and the y-intercept is 8.

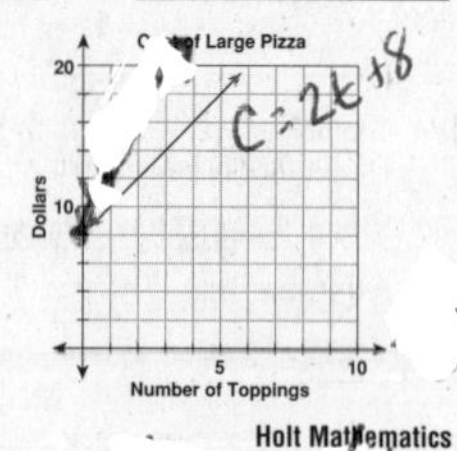

LESSON 12-3 **Practice C**
Using Slopes and Intercepts

Find the x-intercept and y-intercept of each line. Use the intercepts to graph the equation.

1. $3x - 2y = 6$

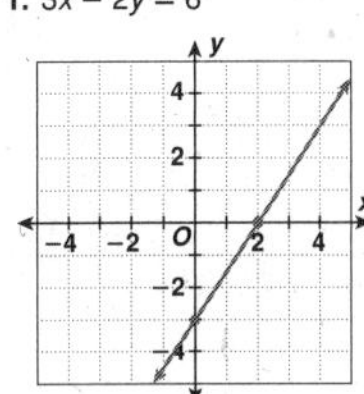

x-intercept is 2; y-intercept is −3

2. $5x + 4y = 20$

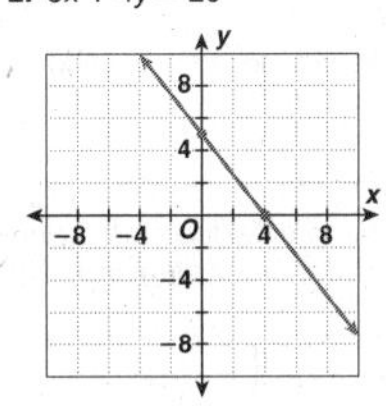

x-intercept is 4; y-intercept is 5

Write each equation in slope-intercept form, and then find the slope and y-intercept.

3. $y - 3x = -10$ — $m = 3;\ b = -10$
4. $3y - 2x = 9$ — $m = \frac{2}{3};\ b = 3$
5. $6y - 2x = -\frac{1}{2}$ — $m = \frac{1}{3};\ b = -\frac{1}{12}$

Write the equation of the line that passes through each pair of points in slope-intercept form.

6. (3, 4), (−1, −4) — $y = 2x - 2$
7. (6, 10), (12, 14) — $y = \frac{2}{3}x + 6$
8. (9, −3), (9, 5) — $x = 9$

9. A home improvement warehouse charges a $60 delivery fee. A customer wants to purchase a number of pieces of lumber that cost $5 a piece. Write an equation in slope-intercept form, where C is the total cost of the delivered lumber and x represents the number of pieces of lumber purchased. Graph the equation for x between 1 and 5 pieces.

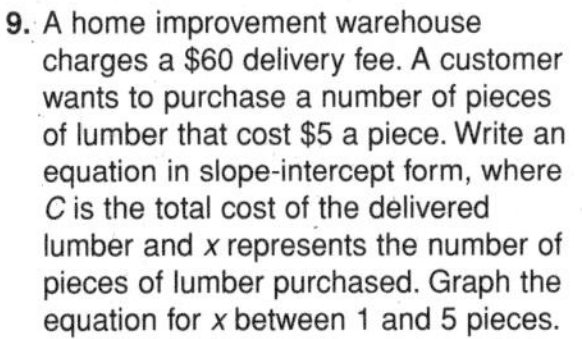
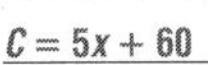

$C = 5x + 60$

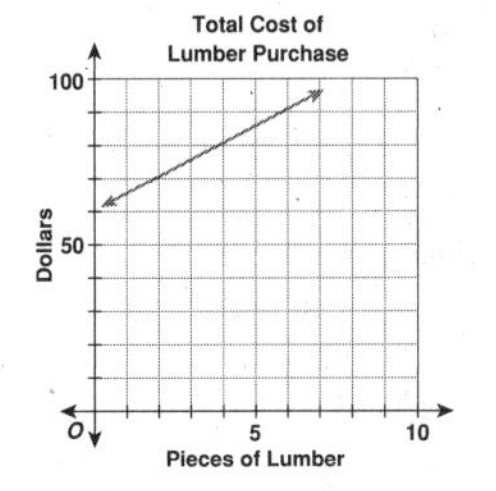

LESSON 12-3 **Reteach**
Using Slopes and Intercepts

x-intercept: the x-coordinate of the point at which a line crosses the x-axis

y-intercept: the y-coordinate of the point at which a line crosses the y-axis

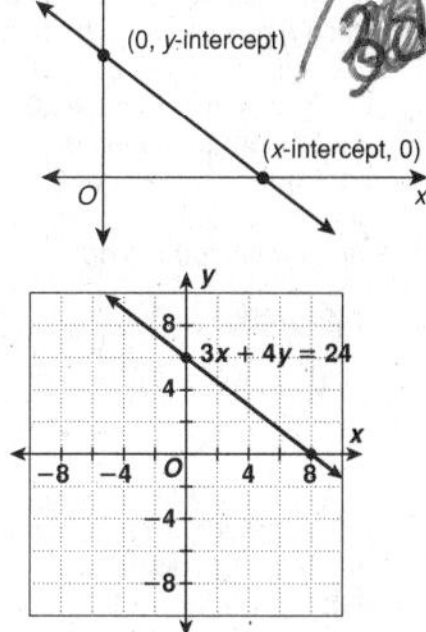

You can find the intercepts of a line from its equation. Then you can use the intercepts to graph the line.

Find the intercepts of the line $3x + 4y = 24$.

For the x-intercept, let $y = 0$.

$3x + 4y = 24$
$3x + 4(0) = 24$
$3x + 0 = 24$
$3x = 24$
$\frac{3x}{3} = \frac{24}{3}$
$x = 8$

The x-intercept is 8.

For the y-intercept, let $x = 0$.

$3x + 4y = 24$
$3(0) + 4y = 24$
$0 + 4y = 24$
$4y = 24$
$\frac{4y}{4} = \frac{24}{4}$
$y = 6$

The y-intercept is 6.

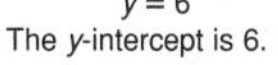

Find the intercepts of each line. Draw both graphs on the same grid.

1. $2x + 3y = 12$

for x-intercept: $2x + 3(\underline{0}) = 12$; $2x = 12$; $\frac{2x}{2} = \frac{12}{2}$; $x = \underline{6}$

for y-intercept: $2(\underline{0}) + 3y = 12$; $3y = 12$; $\frac{3y}{3} = \frac{12}{3}$; $y = \underline{4}$

2. $6y - 3x = 6$

for x-intercept: $6(\underline{0}) - 3x = 6$; $-3x = 6$; $\frac{-3x}{-3} = \frac{6}{-3}$; $x = \underline{-2}$

for y-intercept: $6y - 3(\underline{0}) = 6$; $6y = 6$; $\frac{6y}{6} = \frac{6}{6}$; $y = \underline{1}$

2x + 3y = 12
6y − 3x = 6

LESSON 12-3 **Reteach**
Using Slopes and Intercepts (continued)

slope-intercept form

$y = mx + b$ — slope (m), y-intercept (b)

In this form, the coefficient of x is the slope and the constant term is the y-intercept.

To rewrite an equation in slope-intercept form, *isolate y*.

$2x + 3y = -12$
$-2x \quad -2x$ Subtract $2x$.
$3y = -2x - 12$
$\frac{3y}{3} = \frac{-2}{3}x - \frac{12}{3}$ Divide by 3.
$y = \frac{-2}{3}x - 4$

So, $m = -\frac{2}{3}$ and $b = -4$.

Write each equation in slope-intercept form and then find the slope and y-intercept.

3. $3y = 4x + 15$

$\frac{3y}{3} = \frac{4}{3}x + \frac{15}{3}$ Divide by 3.

$y = \frac{4}{3}x + 5;\ m = \frac{4}{3};\ b = 5$

4. $3x - 2y = 6$

$-3x \quad -3x$ Subtract $3x$.

$y = \frac{3}{2}x - 3;\ m = \frac{3}{2};\ b = -3$

Given two points of a line, you can write its equation.

(2, 5) and (−1, −4)

slope $= \frac{y_2 - y_1}{x_2 - x_1} = \frac{-4 - 5}{-1 - 2} = \frac{-9}{-3} = 3$

To find b, substitute the slope and the values from one of the points into the slope-intercept equation.

$y = mx + b \rightarrow 5 = 3(2) + b$
$5 = 6 + b$
$-1 = b$

So, the equation for the line that passes through (2, 5) and (−1, −4) is $y = 3x - 1$.

Write the equation of the line that passes through each pair of points in slope-intercept form.

5. (2, 11) and (0, 3) — $y = 4x + 3$
6. (−1, 3) and (4, −2) — $y = -x + 2$
7. (10, 1) and (6, −1) — $y = \frac{1}{2}x - 4$

LESSON 12-3 **Challenge**
Another View

The **intercepts** of a line are the points where the line crosses the coordinate axes.

When an equation was in standard form, you found the intercepts by setting one variable and then the other equal to zero.

1. Find the intercepts of the line whose equation is $3x + 5y = 15$.

For the x-intercept, let $y = 0$. For the y-intercept, let $x = 0$.

The x-intercept is $\underline{5}$. The y-intercept is $\underline{3}$.

When an equation is in standard form, you can also find the intercepts by dividing both sides by the constant (on the right side).

2. Divide both sides of the equation $3x + 5y = 15$ by 15, and simplify. Compare the results to those obtained in Question 1.

$\frac{3x}{15} + \frac{5y}{15} = \frac{15}{15};\ \frac{x}{5} + \frac{y}{3} = 1$

The denominators of the variables are the intercepts.

3. Using b to represent the y-intercept and a to represent the x-intercept, write an equation that generalizes the observation you made in Question 2.

$\frac{x}{a} + \frac{y}{b} = 1$

4. a. Using the form of the equation you wrote in Question 3, find the intercepts of the linear equation $2x + 3y = 24$.

$\frac{2x}{24} + \frac{3y}{24} = \frac{24}{24};\ \frac{x}{12} + \frac{y}{8} = 1$

The x-intercept is $\underline{12}$ and the y-intercept is $\underline{8}$.

b. Check your result by using the first method to find the intercepts.

$2x + 3(0) = 24$; $2x = 24$; $x = 12$

$2(0) + 3y = 24$; $3y = 24$; $y = 8$

LESSON 12-3
Problem Solving
Using Slopes and Intercepts

Write the correct answer.

1. Jaime purchased a $20 bus pass. Each time she rides the bus, $1.25 is deducted from the pass. The linear equation $y = -1.25x + 20$ represents the amount of money on the bus pass after x rides. Identify the slope and the x- and y-intercepts. Graph the equation at the right.

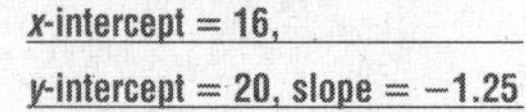

x-intercept = 16,

y-intercept = 20, slope = −1.25

2. The rent charged for space in an office building is related to the size of the space rented. The rent for 600 square feet of floor space is $750, while the rent for 900 square feet is $1150. Write an equation for the rent y based on the square footage of the floor space x.

$y = \frac{4}{3}x - 50$

Choose the letter of the correct answer.

3. A limousine charges $35 plus $2 per mile. Which equation shows the total cost of a ride in the limousine?
   - A $y = 35x + 2$
   - B $y = 2x + 35$
   - C $y = 2x - 35$
   - D $2x + 35y = 2$

4. A newspaper pays its delivery people $75 each day plus $0.10 per paper delivered. Which equation shows the daily earnings of a delivery person?
   - F $y = 0.1x + 75$
   - G $y = 75x + 0.1$
   - H $x + 0.1y = 75$
   - J $0.1x + y = 75$

5. A friend gave Ms. Morris a $50 gift card for a local car wash. If each car wash costs $6, which equation shows the number of dollars left on the card?
   - A $50x + 6y = 1$
   - B $y = 6x + 50$
   - C $y = -6x + 50$
   - D $y = 6x - 50$

6. Antonio's weekly allowance is given by the equation $A = 0.5c + 10$, where c is the number of chores he does. If he received $16 in allowance one week, how many chores did he do?
   - F 10
   - G 12
   - H 14
   - J 15

LESSON 12-3
Reading Strategies
Use a Visual Model

Refer to the coordinate plane at the right. Find the point where the line crosses the x-axis. This point is called the **x-intercept.**

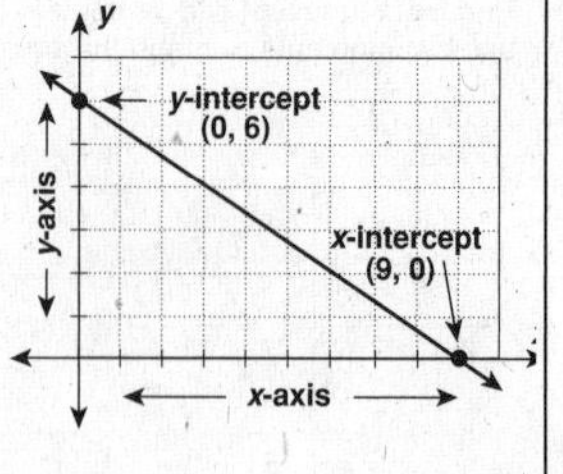

1. What is the y-value of the ordered pair for this point?

0

Find the point where the line crosses the y-axis. This point is called the **y-intercept.**

2. What is the x-value of the ordered pair for this point?

0

3. Which axis does the line cross at the x-intercept?

the x-axis

4. Name the ordered pair for the point where the line crosses the x-axis.

(9, 0)

5. Which axis does the line cross at the y-intercept?

y-axis

6. Name the ordered pair for the point where the line crosses the y-axis.

(0, 6)

LESSON 12-3
Puzzles, Twisters & Teasers
Word Bath!

Circle words from the list in the word search (horizontally, vertically or diagonally). Find a word that answers the riddle.

intercept, axis, slope, point, form, line, graph, rate, coordinate, change

```
C H A N G E L I N E F Q
O L H J U I R N R T O W
O P E E R T Y T Y U R E
R F G A S K I E O M M P
D C V B N A L R P L M O
I W E R T Y U C N K O I
N D F G H J K E B J I N
A X I S S L O P E U H T
T A S D F R A T E V G Y
E G R A P H C X Z A S D
```

Why did the robber take a bath?

Because he wanted to make a clean getaway.

LESSON 12-4
Practice A
Point-Slope Form

Use the point-slope form of each equation to identify the slope of the line.

1. $y - 1 = 2(x - 3)$ — $m = 2$
2. $y + 4 = -1(x + 7)$ — $m = -1$
3. $y - 7 = -3(x + 8)$ — $m = -3$

Use the point-slope form of each equation to identify a point each line passes through.

4. $y - 1 = -2(x - 4)$ — $(x_1, y_1) = (4, 1)$
5. $y + 3 = -5(x - 1)$ — $(x_1, y_1) = (1, -3)$
6. $y + 5 = -2(x + 6)$ — $(x_1, y_1) = (-6, -5)$

Use the point-slope form of each equation to identify a point the line passes through and the slope of the line.

7. $y - 9 = 2(x - 3)$ — $m = 2$; $(x_1, y_1) = (3, 9)$
8. $y + 6 = -3(x - 1)$ — $m = -3$; $(x_1, y_1) = (1, -6)$
9. $y + 1 = -7(x + 2)$ — $m = -7$; $(x_1, y_1) = (-2, -1)$
10. $y + 2 = -6(x - 7)$ — $m = -6$; $(x_1, y_1) = (7, -2)$
11. $y + 6 = -5(x + 9)$ — $m = -5$; $(x_1, y_1) = (-9, -6)$
12. $y - 3 = \frac{1}{3}(x + 9)$ — $m = \frac{1}{3}$; $(x_1, y_1) = (-9, 3)$

Write the point-slope form of the equation with the given slope that passes through the indicated point.

13. the line with slope −4 passing though (5, 4)

$y - 4 = -4(x - 5)$

14. the line with slope 2 passing through (−1, −2)

$y + 2 = 2(x + 1)$

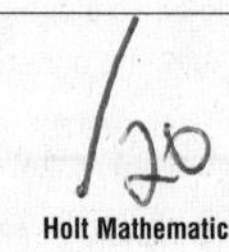

LESSON **12-4**

Practice B

Point-Slope Form

Use the point-slope form of each equation to identify a point the line passes through and the slope of the line.

1. $y - 2 = 4(x - 1)$
$m = 4$; $(x_1, y_1) = (1, 2)$

2. $y + 1 = 2(x - 3)$
$m = 2$; $(x_1, y_1) = (3, -1)$

3. $y - 4 = -3(x + 1)$
$m = -3$; $(x_1, y_1) = (-1, 4)$

4. $y + 5 = -2(x + 6)$
$m = -2$; $(x_1, y_1) = (-6, -5)$

5. $y + 4 = -9(x + 3)$
$m = -9$; $(x_1, y_1) = (-3, -4)$

6. $y - 7 = -7(x - 7)$
$m = -7$; $(x_1, y_1) = (7, 7)$

7. $y - 10 = 6(x - 8)$
$m = 6$; $(x_1, y_1) = (8, 10)$

8. $y + 12 = 2.5(x + 4)$
$m = 2.5$; $(x_1, y_1) = (-4, -12)$

9. $y + 8 = \frac{1}{2}(x - 3)$
$m = \frac{1}{2}$; $(x_1, y_1) = (3, -8)$

Write the point-slope form of the equation with the given slope that passes through the indicated point.

10. the line with slope -1 passing through $(2, 5)$
$y - 5 = -1(x - 2)$

11. the line with slope 2 passing through $(-1, 4)$
$y - 4 = 2(x + 1)$

12. the line with slope 4 passing through $(-3, -2)$
$y + 2 = 4(x + 3)$

13. the line with slope 3 passing through $(7, -6)$
$y + 6 = 3(x - 7)$

14. the line with slope -3 passing through $(-6, 4)$
$y - 4 = -3(x + 6)$

15. the line with slope -2 passing through $(5, 1)$
$y - 1 = -2(x - 5)$

16. Michael was driving at a constant speed of 60 mph when he crossed the Sandy River. After 1 hour, he passed a highway marker for mile 84. Write an equation in point-slope form, and find which highway marker he will pass 90 minutes after crossing the Sandy River.
$y - 84 = 60(x - 1)$; highway marker for 114 miles

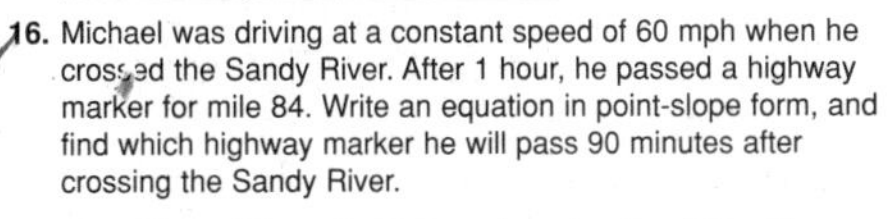

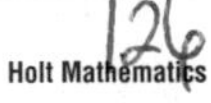

LESSON **12-4**

Practice C

Point-Slope Form

Write the point-slope form of the equation with the given slope that passes through the indicated point.

1. the line with slope $\frac{1}{2}$ passing through $(-4, 8)$
$y - 8 = \frac{1}{2}(x + 4)$

2. the line with slope 7 passing through $(\frac{1}{3}, -6)$
$y + 6 = 7\left(x - \frac{1}{3}\right)$

3. the line with slope 2.6 passing through $(7.8, 4.5)$
$y - 4.5 = 2.6(x - 7.8)$

4. the line with slope $\frac{5}{3}$ passing through $(2, 5)$
$y - 5 = \frac{5}{3}(x - 2)$

5. the line with slope $-\frac{3}{4}$ passing through $(\frac{1}{4}, \frac{1}{5})$
$y - \frac{1}{5} = -\frac{3}{4}\left(x - \frac{1}{4}\right)$

6. the line with slope -9 passing through $(-\frac{2}{3}, -9)$
$y + 9 = -9\left(x + \frac{2}{3}\right)$

The slopes of parallel lines are equal. The slopes of perpendicular lines are negative reciprocals. (If line *A* has a slope of 2 and line *A* is perpendicular to line *B*, then the slope of line *B* is $-\frac{1}{2}$.)

Write the point-slope form of each line described below.

7. the line parallel to $y = 5x - 1$ that passes through $(-2, 7)$
$y - 7 = 5(x + 2)$

8. the line perpendicular to $y = 3x + 6$ that passes through $(-1, 0)$
$y = -\frac{1}{3}(x + 1)$

9. the line perpendicular to $y = -\frac{2}{3}x$ that passes through $(-5, -5)$
$y + 5 = \frac{3}{2}(x + 5)$

10. the line parallel to $y = \frac{3}{4}x + 8$ that passes through $(-1, -9)$
$y + 9 = \frac{3}{4}(x + 1)$

11. A school librarian is packing up books for the summer. The boxes will hold either 6 English books and 18 math books, or 11 English books and 14 math books. Let x equal the number of English books and y equal the number of math books. Write two different equations in point-slope form using this information.
$y - 18 = -\frac{4}{5}(x - 6)$ or $y - 14 = -\frac{4}{5}(x - 11)$

LESSON **12-4**

Reteach

Point-Slope Form

$y - y_1 = m(x - x_1)$

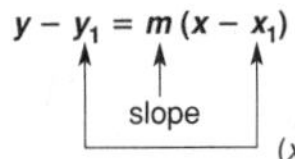

slope

(x_1, y_1) are the coordinates of a known point on the line.

If a minus sign precedes a coordinate value, the coordinate is positive.	If a plus sign precedes a coordinate value, the coordinate is negative.
$y - 3 = 7(x - 1)$	$y + 3 = 7(x + 1)$
$(1, 3)$ is on the line; slope $m = 7$	$(-1, -3)$ is on the line; slope $m = 7$

Identify the slope of each line and a point it passes through.

	1. $y + 2 = 5(x - 3)$	2. $y - 4 = -3(x + 5)$
	$m =$ 5	$m =$ −3
Which sign for each coordinate?	opposite; same	same; opposite
Coordinates of a point on the line:	(3, −2)	(−5, 4)

To write an equation for the line with slope -4 that passes through $(6, -2)$, substitute $m = -4$, $x_1 = 6$, $y_1 = -2$ into the point-slope form.

$y - y_1 = m(x - x_1)$
$y - (-2) = -4(x - 6)$
$y + 2 = -4(x - 6)$

Write the point-slope form of the equation with the given slope that passes through the given point.

3. $m = 3$; $(x_1, y_1) = (7, 2)$
$y - y_1 = m(x - x_1)$
$y -$ 2 $=$ 3 $(x -$ 7 $)$

4. $m = -5$; $(x_1, y_1) = (2, 6)$
$y - y_1 = m(x - x_1)$
$y -$ 6 $=$ −5 $(x -$ 2 $)$

5. $m = \frac{1}{2}$; $(x_1, y_1) = (-8, 1)$
$y - y_1 = m(x - x_1)$
$y -$ 1 $=$ $\frac{1}{2}$ $(x$ + 8 $)$

6. $m = -\frac{3}{4}$; $(x_1, y_1) = (0, -1)$
$y - y_1 = m(x - x_1)$
y + 1 $=$ $-\frac{3}{4}$ $(x -$ 0 $)$

LESSON **12-4**

Challenge

So Everyone Gets the Same Answer

The **standard form** of a line is $Ax + By = C$ where A, B, and C are *real numbers*.

To write an equation of a line, you need to know two pieces of information.

When the slope and the y-intercept are known, use $y = mx + b$.
When the slope and a point on the line are known, use $y - y_1 = m(x - x_1)$.

You can use either the slope-intercept form or the point-slope form to write an equation in standard form.

Write an equation in standard form for the line that contains side $\overline{AB}$ of triangle ABC.

Use $A(1, 0)$ and $B(4, 5)$ to find the slope of $\overleftrightarrow{AB}$. $m = \frac{5 - 0}{4 - 1} = \frac{5}{3}$

Substitute $m = \frac{5}{3}$ and $(x_1, y_1) = (1, 0)$ into point-slope form. $y - 0 = \frac{5}{3}(x - 1)$

Write the equation in standard form.

clear fractions	$3y = 5(x - 1)$
distribute	$3y = 5x - 5$
add and subtract	$5x - 3y = 5$

Write the standard form of the equation for each indicated line.

1. 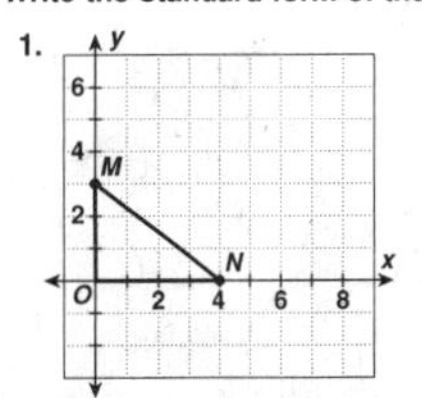

$\overline{MN}$ of right triangle MNO
$3x + 4y = 12$

2. 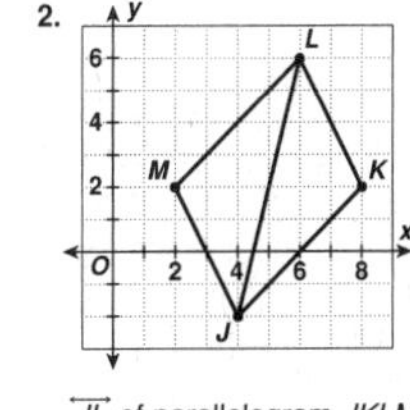

$\overleftrightarrow{JL}$ of parallelogram $JKLM$
$4x - y = 18$

LESSON 12-4
Problem Solving
Point-Slope Form

Write the correct answer.

1. A 1600 square foot home in Houston will sell for about \$102,000. The price increases about \$43.41 per square foot. Write an equation that describes the price y of a house in Houston, based on the square footage x.

   $y - 102,000 = 43.41(x - 1600)$

2. Write the equation in Exercise 1 in slope-intercept form.

   $y = 43.41x + 32,544$

3. Wind chill is a measure of what temperature feels like with the wind. With a 25 mph wind, 40°F will feel like 29°F. Write an equation in point-slope form that describes the wind chill y based on the temperature x, if the slope of the line is 1.337.

   $y - 29 = 1.337(x - 40)$

4. With a 25 mph wind, what does a temperature of 0°F feel like?

   −24.48°F

From 2 to 13 years, the growth rate for children is generally linear. Choose the letter of the correct answer.

5. The average height of a 2-year old boy is 36 inches, and the average growth rate per year is 2.2 inches. Write an equation in point-slope form that describes the height of a boy y based on his age x.

   A $y - 36 = 2(x - 2.2)$
   B $y - 2 = 2.2(x - 36)$
   Ⓒ $y - 36 = 2.2(x - 2)$
   D $y - 2.2 = 2(x - 36)$

6. The average height of a 5-year old girl is 44 inches, and the average growth rate per year is 2.4 inches. Write an equation in point-slope form that describes the height of a girl y based on her age x.

   F $y - 2.4 = 44(x - 5)$
   Ⓖ $y - 44 = 2.4(x - 5)$
   H $y - 44 = 5(x - 2.4)$
   J $y - 5 = 2.4(x - 44)$

7. Write the equation from Exercise 6 in slope-intercept form.

   A $y = 2.4x - 100.6$
   B $y = 44x - 217.6$
   C $y = 5x + 32$
   Ⓓ $y = 2.4x + 32$

8. Use the equation in Exercise 6 to find the average height of a 13-year old girl.

   F 56.3 in.
   Ⓖ 63.2 in.
   H 69.4 in.
   J 97 in.

LESSON 12-4
Reading Strategies
Use a Procedure

To find the slope of a line, you can use the coordinates for two points on the line.

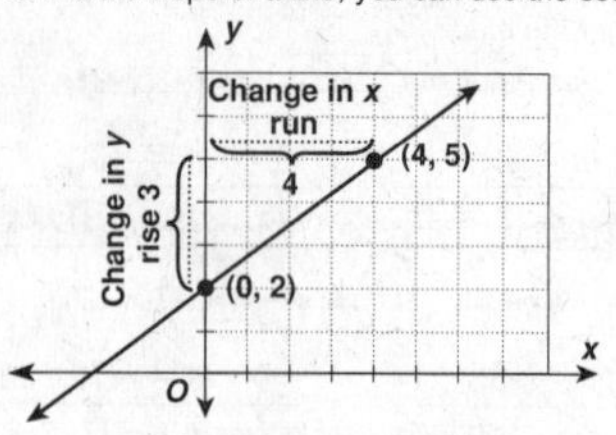

Step 1: Subtract to find the difference between the y-coordinates of the two points: $5 - 2 = 3$.

Step 2: Subtract to find the difference between the x-coordinates of the two points: $4 - 0 = 4$.

Step 3: Write the ratio of the differences. The difference between the y-coordinates is 3. The difference between the x-coordinates is 4. The slope is the ratio $\frac{3}{4}$.

When you have the slope of a line and a point it passes through, you can write an equation for the line in **point-slope form.**

Point on the line	Slope	Point-slope form
(x_1, y_1)	m	$y - y_1 = m(x - x_1)$
$(4, 5)$	$\frac{3}{4}$	$y - 5 = \frac{3}{4}(x - 4)$

Use the above example to answer each question.

1. What ratio is shown by the slope of a line?

   $\frac{\text{change in } y}{\text{change in } x}$, or $\frac{\text{rise}}{\text{run}}$

2. How can you find the slope of a line?

   Possible answer: Find the difference between the y-values of two points and find the difference between the x-values of the same points; write the differences as a ratio.

3. What information do you need to write an equation for a line in point-slope form?

   coordinates of a point the line passes through and slope of the line

LESSON 12-4
Puzzles, Twisters & Teasers
Get a Clue!

Identify a point on each line and the slope of the line. Then use the slope values to answer the riddle.

1. $y + 1 = \frac{2}{3}(x + 7)$ point = −7, −1 slope = $\frac{2}{3}$ S
2. $y + 1 = 11(x - 1)$ point = 1, −1 slope = 11 C
3. $y - 2 = -\frac{1}{6}(x - 11)$ point = 11, 2 slope = $-\frac{1}{6}$ N
4. $y + 7 = 1(x - 5)$ point = 5, −7 slope = 1 L
5. $y + 7 = 3(x + 4)$ point = −4, −7 slope = 3 E
6. $y - 9 = 5(x - 12)$ point = 12, 9 slope = 5 B
7. $y - 11 = 14(x - 8)$ point = 8, 11 slope = 14 H
8. $y - 4 = -2(x + 7)$ point = −7, 4 slope = −2 O
9. $y - 3 = -1.8(x - 5.6)$ point = 5.6, 3 slope = −1.8 R
10. $y + 8 = -6(x - 9)$ point = 9, −8 slope = −6 K

What do you call a dog detective?

S	H	E	R	L	O	C	K
$\frac{2}{3}$	14	3	−1.8	1	−2	11	−6

B	O	N	E	S
5	−2	$-\frac{1}{6}$	3	$\frac{2}{3}$

LESSON 12-5
Practice A
Direct Variation

The following tables show direct variation for the given equation. Complete the missing information in the tables.

1. $y = 2x$

x	−10	−7	−4	3	6	12	15	22
y	−20	−14	−8	6	12	24	30	44

2. $y = \frac{1}{3}x$

x	−21	−15	−9	3	8	12	19	30
y	−7	−5	−3	1	$2\frac{2}{3}$	4	$6\frac{1}{3}$	10

3. Make a graph to determine whether the data sets show direct variation.

x	y
−8	−4
−6	−3
0	0
2	1
4	2
6	3

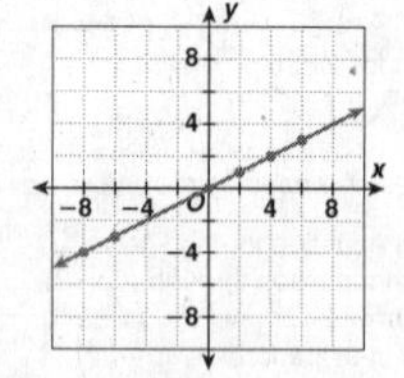

The data sets show direct variation.

Find each equation of direct variation, given that y varies directly with x.

4. y is 10 when x is 2.

   $y = 5x$

5. y is 42 when x is −6.

   $y = -7x$

6. y is −50 when x is 5.

   $y = -10x$

7. y is 15 when x is 30.

   $y = \frac{1}{2}x$

8. At a constant speed, the gasoline a car uses varies directly with the distance the car travels. A car uses 10 gallons of gasoline to travel 210 miles. How many gallons will the car use to travel 294 miles? 14 gallons

LESSON 12-5

Practice B

Direct Variation

Make a graph to determine whether the data sets show direct variation.

1.

x	y
6	9
4	6
0	0
−2	−3
−8	−12

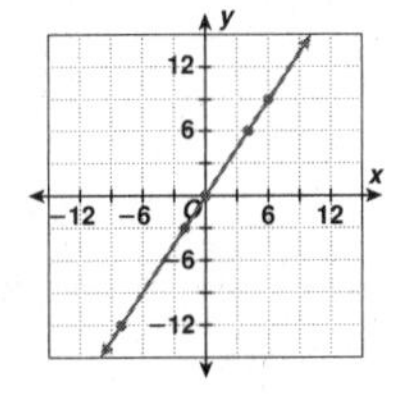

The data sets show direct variation.

2. Write the equation of direct variation for Exercise 1.

$y = 1.5x$ or $y = \frac{3}{2}x$

Find each equation of direct variation, given that y varies with x.

3. y is 32 when x is 4

$y = 8x$

4. y is −10 when x is −20

$y = \frac{1}{2}x$

5. y is 63 when x is −7

$y = -9x$

6. y is 40 when x is 50

$y = \frac{4}{5}x$

7. y is 87.5 when x is 25

$y = 3.5x$

8. y is 90 when x is 270

$y = \frac{1}{3}x$

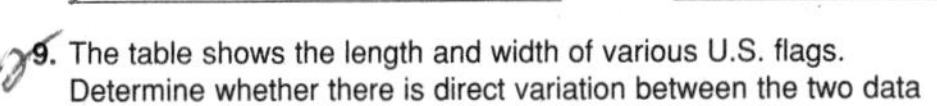

9. The table shows the length and width of various U.S. flags. Determine whether there is direct variation between the two data sets. If so, find the equation of direct variation.

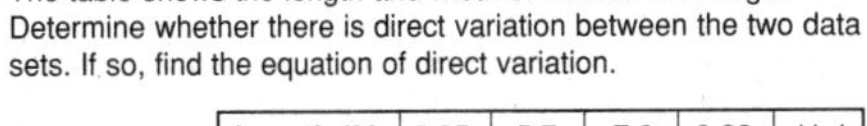

Length (ft)	2.85	5.7	7.6	9.88	11.4
Width (ft)	1.5	3	4	5.2	6

There is direct variation between the lengths and widths of the flags.

$y = 1.9x$, where y is the length, x is the width, and 1.9 is the constant of proportionality

LESSON 12-5

Practice C

Direct Variation

Find each equation of direct variation, given that y varies directly with x.

1. y is 189 when x is 45

$y = 4.2x$

2. y is 456 when x is 3800

$y = 0.12x$

3. y is 763 when x is 981

$y = \frac{7}{9}x$

4. y is $171\frac{3}{4}$ when x is 916

$y = \frac{3}{16}x$

Tell whether each equation represents direct variation between x and y.

5. $y = \frac{9}{10}x$ — yes

6. $y = xy - 8$ — no

7. $-5x - y = 0$ — yes

8. $y = \frac{24}{x}$ — no

9. $\frac{y}{x} = 8.25$ — yes

10. $x - y = -10$ — no

11. $x = y$ — yes

12. $\frac{1}{3}y = x$ — yes

13. The following table shows the distance on a map in inches x and the actual distance between two cities in miles, y. Determine whether there is direct variation between the two data sets. If so, find the equation of direct variation.

x	$456\frac{1}{4}$	$3\frac{1}{2}$	4	5	$7\frac{1}{4}$	8	$9\frac{1}{8}$	11
y	75	175	200	350	$362\frac{1}{2}$	400	$456\frac{1}{4}$	550

There is no direct variation.

14. A person's weight on Earth varies directly with a person's estimated weight on Venus. If a person weighs 110 pounds on Earth, he or she would weigh an estimated 99.7 pounds on Venus. If a person weighs 125 pounds on Earth, what would be his or her estimated weight to the nearest tenth of a pound on Venus?

113.3 pounds

LESSON 12-5

Reteach

Direct Variation

Two data sets have **direct variation** if they are related by a constant ratio, the **constant of proportionality.** A graph of the data sets is linear and passes through (0, 0).

$y = kx$ **equation of direct variation,** where k is the constant ratio

To determine whether two data sets have direct variation, you can compare ratios. You can also graph the data sets on a coordinate grid.

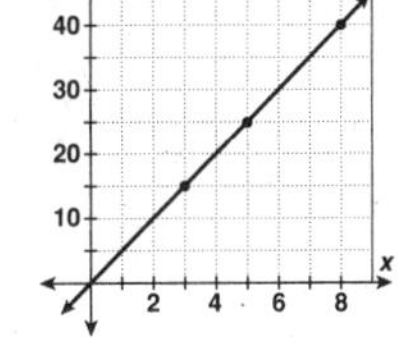
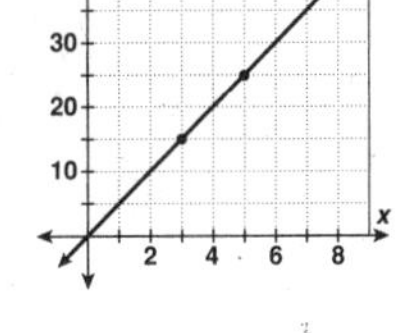
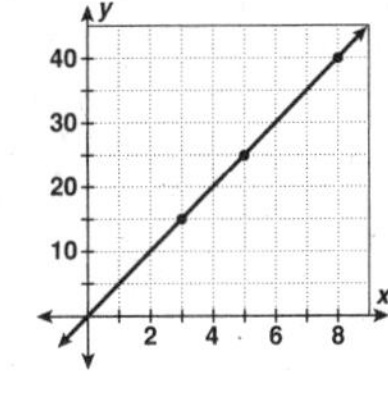
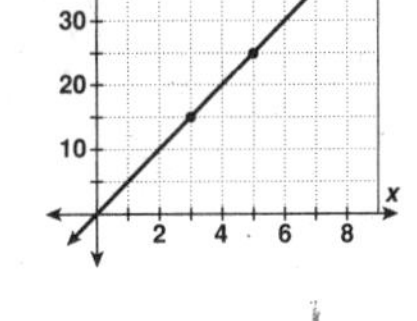
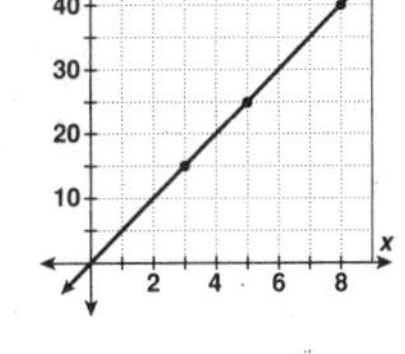
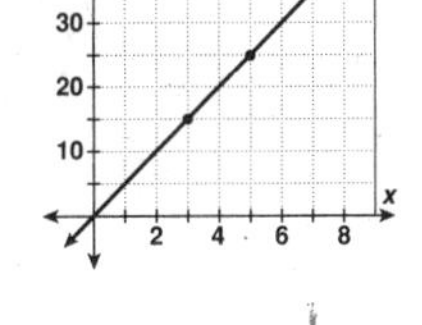
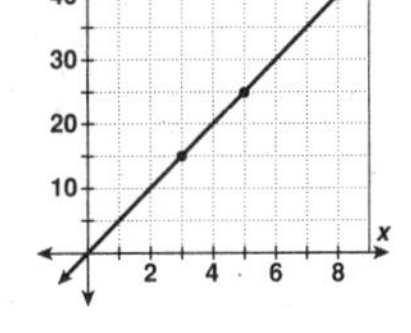
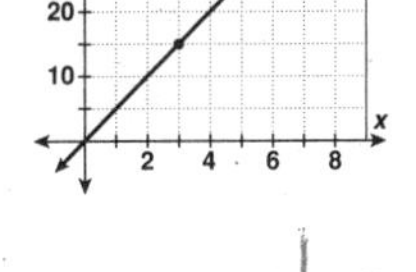
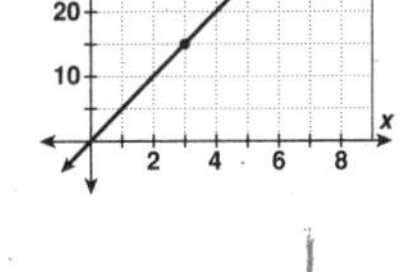
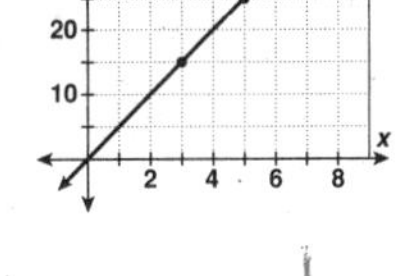
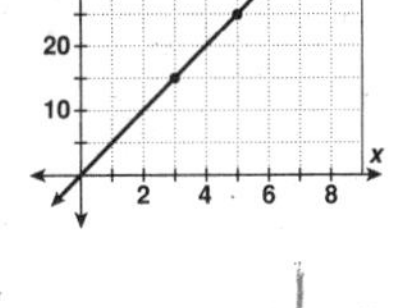
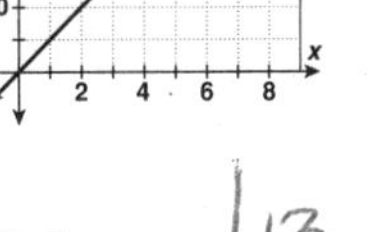
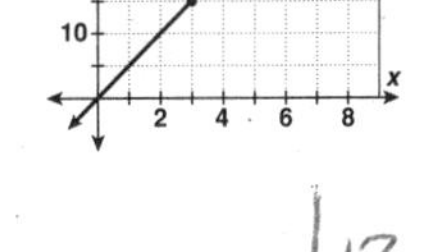

x	3	5	8
y	15	25	40

$\frac{y}{x} = \frac{15}{3} = \frac{25}{5} = \frac{40}{8} = \frac{5}{1}$ ← constant ratio

$k = 5 \rightarrow y = 5x$

The graph of the data sets is linear and passes through (0, 0).

So, the data sets show direct variation.

Determine whether the data sets show direct variation. If there is a constant ratio, identify it and write the equation of direct variation. Plot the points and tell whether the graph is linear.

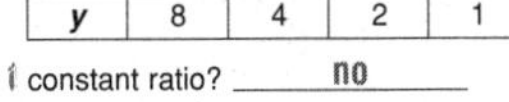
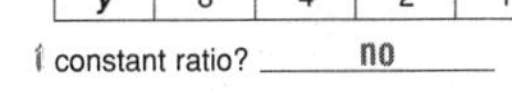
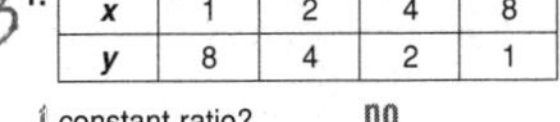
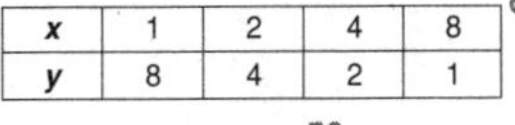
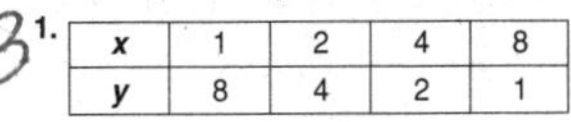
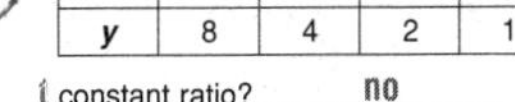

1.

x	1	2	4	8
y	8	4	2	1

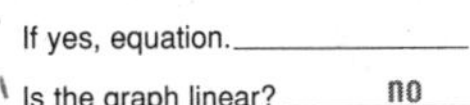

constant ratio? no

If yes, equation.

Is the graph linear? no

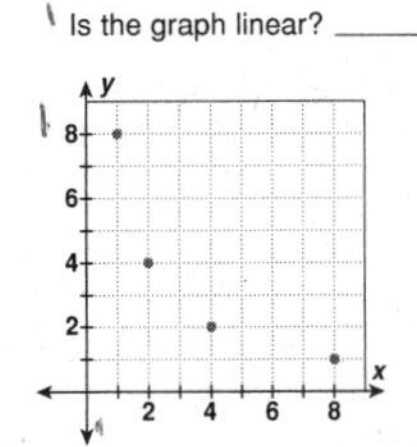

2.

x	0	2	3	5
y	0	20	30	50

constant ratio? yes, 10

If yes, equation. $y = 10x$

Is the graph linear? yes

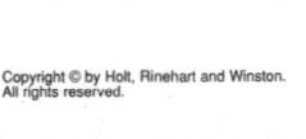

LESSON 12-5

Challenge

Different Paths, Same Result

Problems of direct variation can be solved with two methods.
If r varies directly with h, and $r = 13.5$ when $h = 3$, find r when $h = 7$.

Method 1: Find the constant of variation.

$\frac{r}{h} = k$

$\frac{13.5}{3} = k$ Use a pair of known values.

$4.5 = k$ constant of variation

$r = 4.5h$ equation of variation

$r = 4.5(7) = 31.5$

So, when $h = 7$, $r = 31.5$.

Method 2: Write a proportion.

$\frac{r_1}{h_1} = \frac{r_2}{h_2}$

$\frac{13.5}{3} = \frac{r_2}{7}$ Use all known values.

$3r_2 = 13.5(7)$ Cross multiply.

$\frac{3r_2}{3} = \frac{94.5}{3}$

$r_2 = 31.5$

Use both methods to solve each problem.

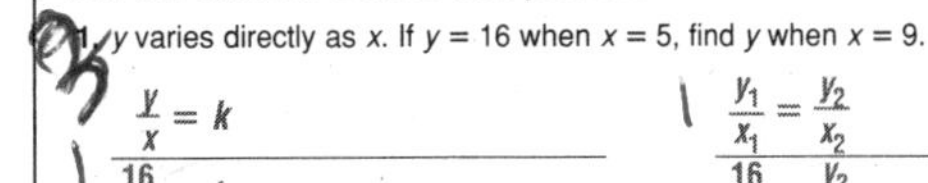

1. y varies directly as x. If $y = 16$ when $x = 5$, find y when $x = 9$.

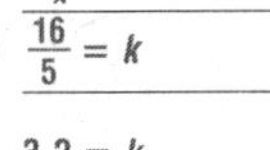

$\frac{y}{x} = k$ $\quad$ $\frac{y_1}{x_1} = \frac{y_2}{x_2}$

$\frac{16}{5} = k$ $\quad$ $\frac{16}{5} = \frac{y_2}{9}$

$3.2 = k$ $\quad$ $5y_2 = 16(9)$

$y = 3.2x$ $\quad$ $\frac{5y_2}{5} = \frac{144}{5}$

$y = 3.2(9) = 28.8$ $\quad$ $y_2 = 28.8$

So, when $x = 9$, $y =$ 28.8.

2. A varies directly as s^2. If $A = 75$ when $s = 5$, find A when $s = 7$.

$\frac{A}{s^2} = k$ $\quad$ $\frac{A_1}{(s_1)^2} = \frac{A_2}{(s_2)^2}$

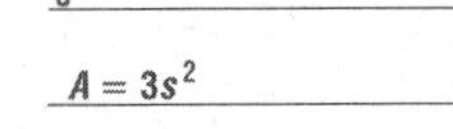

$\frac{75}{5^2} = k; k = 3$ $\quad$ $\frac{75}{5^2} = \frac{A_2}{7^2}$; $25A_2 = 75(49)$

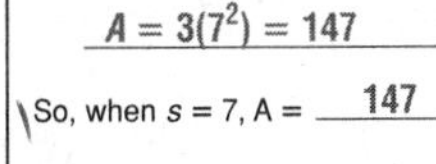

$A = 3s^2$ $\quad$ $\frac{25A_2}{25} = \frac{3675}{25}$

$A = 3(7^2) = 147$ $\quad$ $A_2 = 147$

So, when $s = 7$, $A =$ 147.

LESSON 12-5

Problem Solving
Direct Variation

Determine whether the data sets show direct variation. If so, find the equation of direct variation.

1. The table shows the distance in feet traveled by a falling object in certain times.

Time (s)	0	0.5	1	1.5	2	2.5	3
Distance (ft)	0	4	16	36	64	100	144

No direct variation

2. The R-value of insulation gives the material's resistance to heat flow. The table shows the R-value for different thicknesses of fiberglass insulation.

Thickness (in)	1	2	3	4	5	6
R-value	3.14	6.28	9.42	12.56	15.7	18.84

Direct variation; $R = 3.14t$

3. The table shows the lifting power of hot air.

Hot Air (ft^3)	50	100	500	1000	2000	3000
Lift (lb)	1	2	10	20	40	60

Direct variation; $L = \left(\frac{1}{50}\right)H$

4. The table shows the relationship between degrees Celsius and degrees Fahrenheit.

° Celsius	−10	−5	0	5	10	20	30
° Fahrenheit	14	23	32	41	50	68	86

No direct variation

The relationship between your weight on Earth and your weight on other planets is direct variation. The table below shows how much a person who weights 100 lb on Earth would weigh on the moon and different planets.

Solar System Objects	Weight (lb)
Moon	16.6
Jupiter	236.4
Pluto	6.7

5. Find the equation of direct variation for the weight on earth e and on the moon m.
   (A) $m = 0.166e$ C $m = 6.02e$
   B $m = 16.6e$ D $m = 1660e$

6. How much would a 150 lb person weigh on Jupiter?
   F 63.5 lb   (H) 354.6 lb
   G 286.4 lb   J 483.7 lb

7. How much would a 150 lb person weigh on Pluto?
   A 5.8 lb   C 12.3 lb
   (B) 10.05 lb   D 2238.8 lb

LESSON 12-5

Reading Strategies
Use Tables and Graphs

When quantities are related proportionally by a constant multiplier, they have **direct variation.**

This table shows the relationship between the number of glasses filled and the amount of juice needed to fill them. The amount of juice needed *varies directly* with the number of glasses filled.

Glasses	1	2	3	4
Juice Needed	8 oz	16 oz	24 oz	32 oz

1. What are the quantities that form this direct variation?
   number of glasses filled and ounces of juice needed
2. What is the constant multiplier?
   8

A graph of a direct variation is always linear and always passes through (0,0).

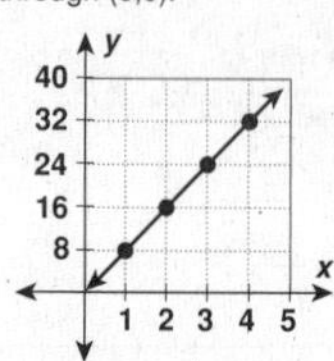

3. What do the x-values on the graph stand for?
   the number of glasses filled
4. What do the y-values on the graph stand for?
   the amount of juice needed for each glass
5. What does the ordered pair (2, 16) mean?
   2 glasses need 16 ounces of juice
6. Write an ordered pair for 3 glasses and the amount of juice needed.
   (3, 24)

LESSON 12-5

Puzzles, Twisters & Teasers
It Just Doesn't Hold Water!

Circle words from the list in the word search (horizontally, vertically or diagonally). You will also find a word that answers the riddle.

direct	variation	constant	proportionality	ratio
graph	quantity	algebra	table	data

```
P R O P O R T I O N A L I T Y
V A C V B N A I K L L Q W E R
A T D F G H B U J M G R A P H
R I E R T Y L T H N E Q T Y U
I O B G H U E W D V B U I O P
A A S D F G H J K L R A A S D
T S I E V E X D A T A N F G H
I Z X C V B N M K I J T K L Q
O D I R E C T H U J I I Z X C
N C O N S T A N T D F T V B N
L K J H G F D S A T B Y M K I
```

What is as round as a dishpan and as deep as a tub, yet the ocean could not fill it?

A sieve.

LESSON 12-6

Practice A
Graphing Inequalities in Two Variables

1. The graph shows $y = x + 2$. Shade the the inequality $y \le x + 2$.

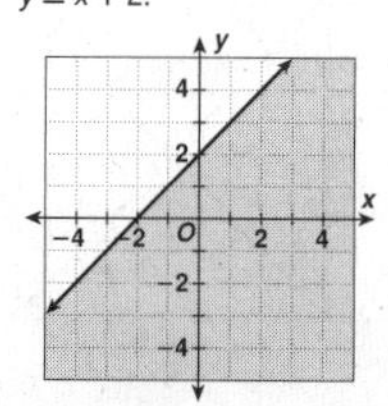

2. The graph shows $y = x$. Shade the side of the line to show the inequality $y \ge x$.

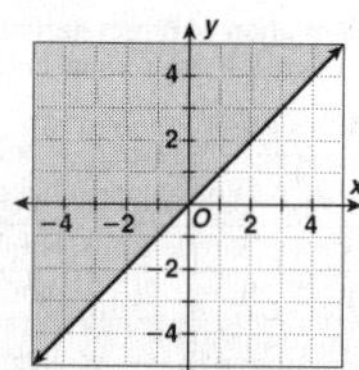

Solve each inequality for y.

3. $x + y \ge 8$ $y \ge -x + 8$
4. $3x - y > 6$ $y < 3x - 6$
5. $x - y \ge -2$ $y \le x + 2$
6. $4x + y \le -3$ $y \le -4x - 3$

Graph each inequality.

7. $y \le x - 1$

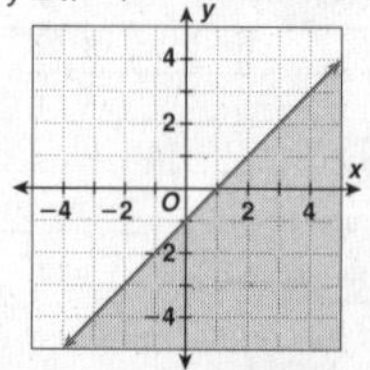

8. $y \ge 2x + 1$

9. Georgia has scored at least 12 points in each of her basketball games this year. She has scored both 2-point and 3-point field goals in every game. Let x equal the 2-point shots and y equal the 3-point shots. Write and graph an inequality to show the number of points she scored each game.
   $2x + 3y \ge 12$, or $y \ge -\frac{2}{3}x + 4$

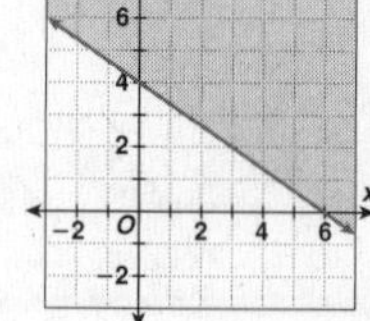

LESSON 12-6
Practice B
Graphing Inequalities in Two Variables

Graph each inequality.

1. $y \geq 2x + 3$

2. $y - 4x \leq 1$

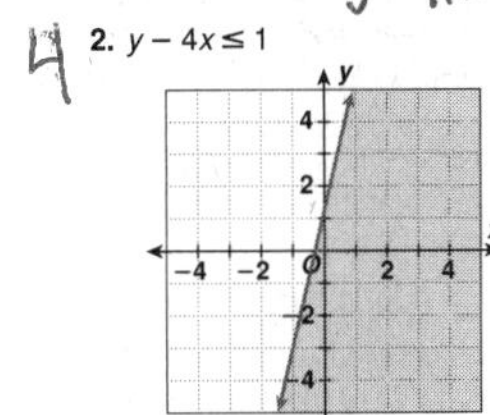

3. $2(3x - y) > 6$

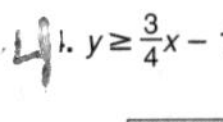

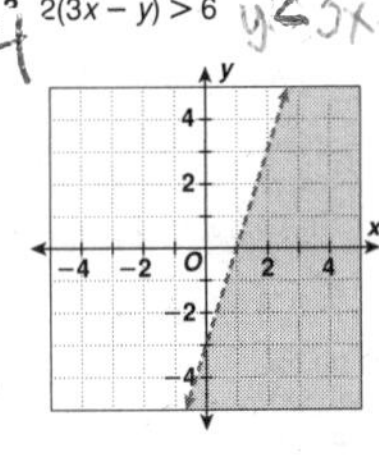

4. $y \geq \frac{3}{4}x - 1$

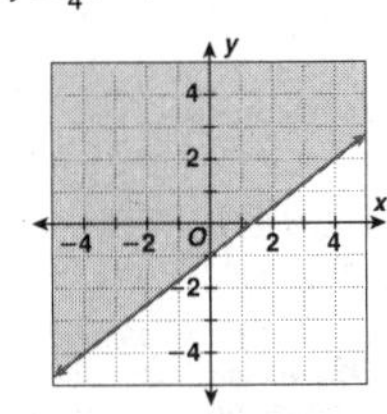

5. a. A theater club hopes to raise at least \$550 on the opening night of its new show. Student tickets for the show cost \$2.75, and adult tickets cost \$5.50. Write and graph an inequality showing the numbers of tickets that would meet the club's goal.

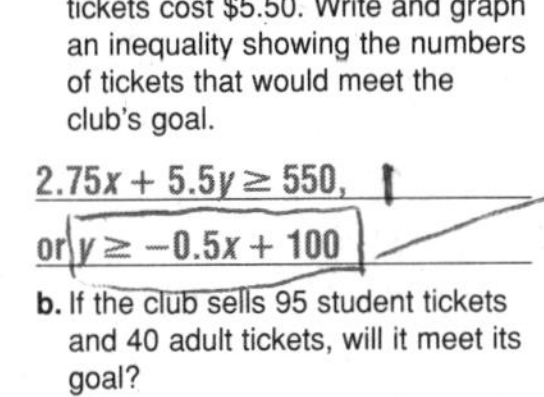

$2.75x + 5.5y \geq 550$, or $y \geq -0.5x + 100$

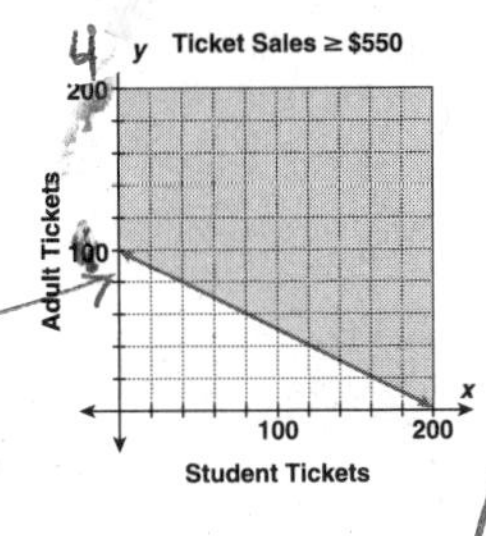

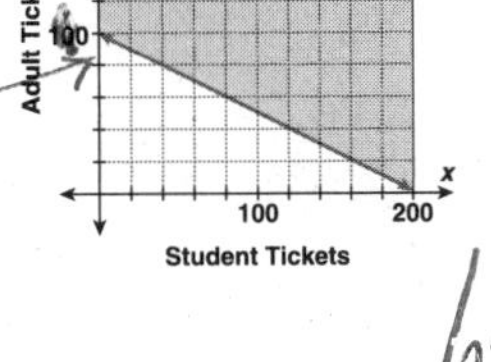

b. If the club sells 95 student tickets and 40 adult tickets, will it meet its goal?

no

LESSON 12-6
Practice C
Graphing Inequalities in Two Variables

Tell whether the given ordered pair is a solution of each inequality shown.

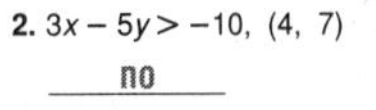

1. $2x + 3y \geq 6$, (2, 1) — yes
2. $3x - 5y > -10$, (4, 7) — no
3. $x - 2y \geq 8$, (5, −2) — yes
4. $5x - 4y \geq -9$, (−3, −2) — yes
5. $y > \frac{1}{2}x + \frac{3}{4}$, (−2, 1) — yes
6. $y \geq \frac{2}{3}x - 8$, (6, −7) — no

7. a. Graph the inequality $3x - 2y \geq 4$.

b. Name an ordered pair that is a solution of the inequality.

Possible answer: (2, −2)

c. Is $\left(6, \frac{1}{2}\right)$ a solution of $3x - 2y \geq 4$?

yes

d. Explain how to check the answer for c.

Possible answer: Substitute 6 for x and $\frac{1}{2}$ for y in the inequality. If the statement is true, then the point is a solution.

e. Which side of the line $3x - 2y = 4$ is shaded?

below the line, or the side that includes $(6, \frac{1}{2})$.

f. Name an ordered pair that is a solution of $3x - 2y \leq 4$.

Possible answer: (0, 0)

8. Jaime is having a party and has budgeted \$120 for the party. He plans on serving pizzas that cost \$6 each and soft drinks that cost \$0.75 each. Let x represent the number of pizzas and y represent the number of soft drinks. Write an inequality showing the numbers of pizzas and soft drinks he could buy while staying within his budget.

$6x + 0.75y \leq 120$, or $y \leq -8x + 160$

9. If Jaime invites 25 people and plans on each person having 2 soft drinks and $\frac{1}{2}$ pizza, will he have budgeted enough money? yes

LESSON 12-6
Reteach
Graphing Inequalities in Two Variables

A **boundary line** divides the coordinate plane into two *half-planes.*

When $y = -x + 4$ is a boundary line:

All the points on the line satisfy the equation.
(5, −1) is on the line since $-1 = -5 + 4$.

All the points in the half-plane above the line satisfy the linear inequality $y > -x + 4$.
(5, 3) is in the half-plane above the line since $3 > -5 + 4$.

All the points in the half-plane below the line satisfy the linear inequality $y < -x + 4$.
(−2, 1) lies in the half-plane below the line since $1 < -(-2) + 4$.

When a boundary line is in the form $y = mx + b$, points in the half-plane above the line satisfy the inequality $y > mx + b$ and points in the half-plane below the line satisfy the inequality $y < mx + b$.

Complete the linear inequality that each point satisfies.

1. (1, −2) is in the half-plane below the boundary line $y = 3x - 4$.
The boundary line is in the form $y = mx + b$; so, (1, −2) satisfies the linear inequality y < $3x - 4$.

2. (−3, 7) is in the half-plane below the boundary line $y = -2x + 6$.
The boundary line is in the form $y = mx + b$; so, (−3, 7) satisfies the linear inequality y < $-2x + 6$.

Write the linear inequality whose solution set is shaded on each graph. The dashed boundary line is not included in the solution set.

3.

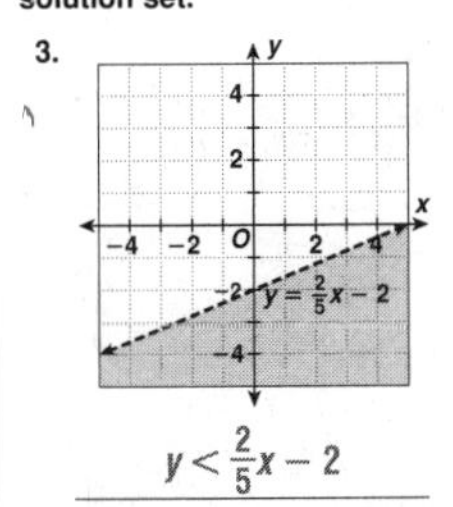

$y < \frac{2}{5}x - 2$

4. $y > -\frac{2}{3}x + 4$

LESSON 12-6
Reteach
Graphing Inequalities in Two Variables (continued)

To graph the solution set of a linear inequality:

Write the equation of the boundary line in the form $y = mx + b$.
Graph the boundary line.
Use a dashed line and shade the region above if $y > mx + b$.
Use a dashed line and shade the region below if $y < mx + b$.
Use a solid line and shade the region above if $y \geq mx + b$.
Use a solid line and shade the region below if $y \leq mx + b$.

Graph the inequality $2x + y \leq -6$:

Write the equation of the boundary line in $y = mx + b$ form.

$2x + y = -6$
$-2x \quad -2x$
$y = -2x - 6$
slope = −2, y-intercept = −6

Since the inequality uses ≤, draw a solid boundary line and shade the half-plane below.

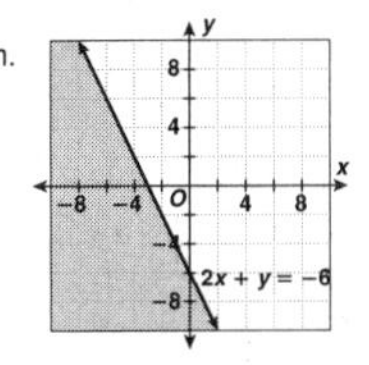

Graph the solution set of each inequality.

5. $3x + y \geq 4$
Rewrite equation $3x + y = 4$.
$y = -3x + 4$
slope = −3, y-intercept = 4
Given symbol is ≥, so draw boundary line solid and shade half-plane above.

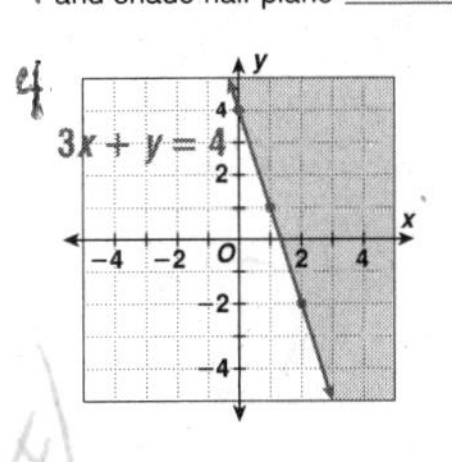

6. $2y - x < 6$
Rewrite equation $2y - x = 6$.
$y = \frac{1}{2}x + 3$
slope = $\frac{1}{2}$, y-intercept = 3
Given symbol is <, so draw boundary line dashed and shade half-plane below.

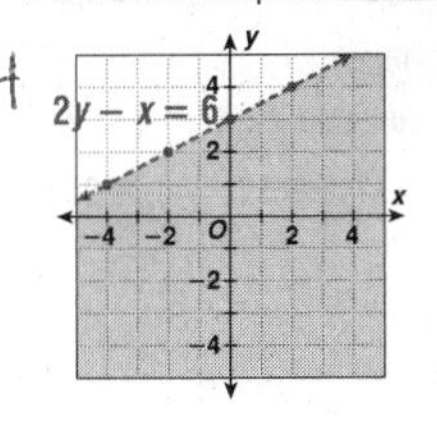

LESSON 12-6
Challenge
Two at a Time

To solve a system of two linear inequalities graphically find the solution set for each linear inequality and mark off the part that overlaps.

Graph the solution set of the system: $x + y \geq 1$, $y < x - 3$

Work with the first inequality.
Rewrite $x + y = 1$ as $y = -x + 1$.
Since line is now in $y = mx + b$ form and given symbol is $\geq$, draw solid boundary line and shade half-plane above line.

Work with the second inequality.
Since boundary line $y = x - 3$ is in $y = mx + b$ form and given symbol is $<$, draw dashed boundary line and shade half-plane below the line.

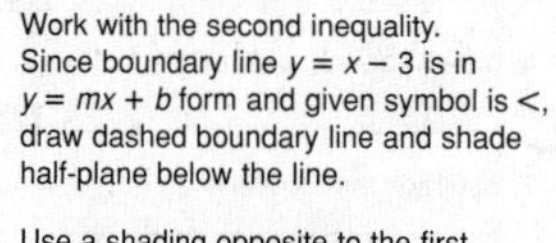

Use a shading opposite to the first shading so that overlap is visible.

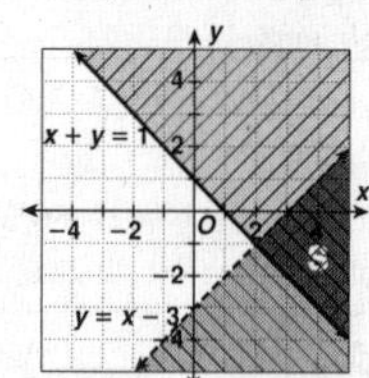

So, the solution set to the system is the cross-hatched region labeled *S*.

Graph the solution set *S* for each system.

1. $y \geq 3x + 1$
   $y < x + 1$

2. $2x - y \leq 4$
   $3x + y < 6$

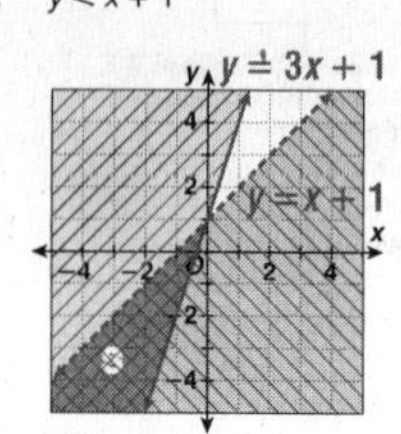

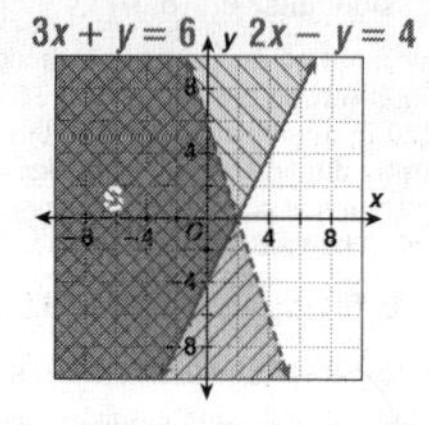

LESSON 12-6
Problem Solving
Graphing Inequalities in Two Variables

The senior class is raising money by selling popcorn and soft drinks. They make $0.25 profit on each soft drink sold, and $0.50 on each bag of popcorn. Their goal is to make at least $500.

1. Write an inequality showing the relationship between the sales of *x* soft drinks and *y* bags of popcorn and the profit goal.

   $0.25x + 0.5y \geq 500$

2. Graph the inequality from exercise 1.

3. List three ordered pairs that represent a profit of exactly $500.

   Possible answers: (800, 600), (400, 800), (1600, 200).

4. List three ordered pairs that represent a profit of more than $500.

   Possible answers: (400, 900), (800, 700), (1600, 300).

5. List three ordered pairs that represent a profit of less than $500.

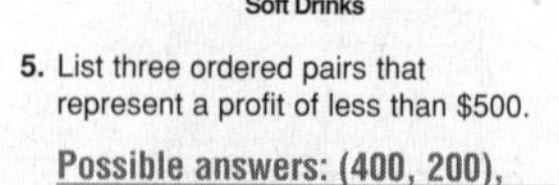

   Possible answers: (400, 200), (800, 400), (1200, 100).

A vehicle is rated to get 19 mpg in the city and 25 mpg on the highway. The vehicle has a 15-gallon gas tank. The graph below shows the number of miles you can drive using no more than 15 gallons.

6. Write the inequality represented by the graph.

   A $\frac{x}{19} + \frac{y}{25} < 15$
   (B) $\frac{x}{19} + \frac{y}{25} \leq 15$
   C $\frac{x}{19} + \frac{y}{25} \geq 15$
   D $\frac{x}{19} + \frac{y}{25} > 15$

7. Which ordered pair represents city and highway miles that you can drive on one tank of gas?

   F (200, 150) H (250, 75)
   G (50, 350) (J) (100, 175)

8. Which ordered pair represents city and highway miles that you cannot drive on one tank of gas?

   A (100, 200) C (50, 275)
   (B) (150, 200) D (250, 25)

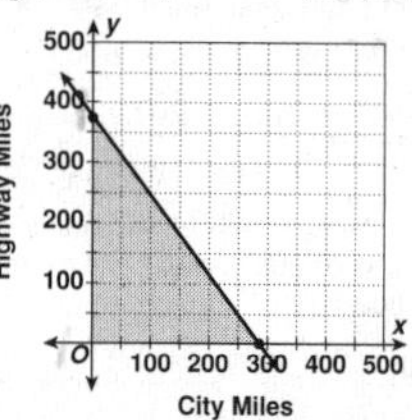

LESSON 12-6
Reading Strategies
Use a Graphic Aid

The graph of a **linear inequality** is half the coordinate plane. The boundary line is the graph of a linear equation.

Linear inequality → $x + y < 10$

Boundry Line → $x + y = 10$

Follow these steps to graph the linear inequality $x + y < 10$.

Step 1: Graph the linear equation showing the boundary line.

x	y
3	7
5	5

Draw a dashed line because none of the points on the line are part of the solution.

Step 2: Test a point on each side of the line.

(2, 0) (0, 11)
$2 + 0 < 10$ $0 + 11 < 10$
$2 < 10$ *true* $11 < 10$ *not true*

Step 3: Shade the side of the line that contains (2, 0).

For 1–4, refer to the example above.

1. What is the first step in graphing a linear inequality?

   graph the linear equation for the boundary line

2. Why is the line dashed instead of solid?

   because the points on the line are not part of the solution

3. What is the second step in graphing a linear inequality?

   test points on either side of the line

4. What does the shaded region of the graph show?

   points that make the inequality true

LESSON 12-6
Puzzles, Twisters & Teasers
Dog-Food for Thought!

Determine whether the given ordered pair is a solution of the inequality. Circle the letter of the correct answer. Then use the letters above the answers to solve the riddle.

1. $y > x + 1$ (0, 0) — X solution; (D) not a solution
2. $y \leq x + 1$ (2, 1) — (O) solution; Q not a solution
3. $3y + 4x \leq 12$ (0, 0) — (G) solution; Z not a solution
4. $y < 330x + 0$ (5, 0) — (B) solution; N not a solution
5. $y \leq 2x + 4$ (−7, 5) — Y solution; (I) not a solution
6. $y > -x + 12$ (−3, 9) — D solution; (S) not a solution

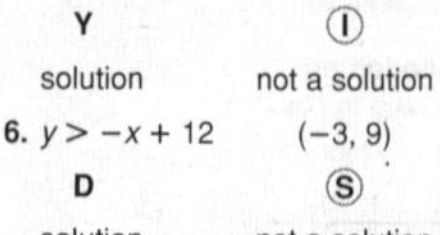

7. $y > -6x + 1$ (2, −14) — H solution; (C) not a solution
8. $y > 3x - 3$ (3, 3) — G solution; (U) not a solution
9. $y < 3(x - 2)$ (5, 1) — (I) solution; R not a solution
10. $y \leq 7(x - 3)$ (5, 14) — (T) solution; P not a solution
11. $y < -1.5x + 2.5$ (−1, 6) — L solution; (S) not a solution

What do you get if you cross a beagle with some bread dough?

dog biscuits

LESSON **12-7**

Practice A

Lines of Best Fit

Find the mean of the *x*- and *y*-coordinates of each set of data.

1.

x	1	2	4	5	7	3	6	8
y	5	6	7	8	10	7	10	11

$x_m = 4.5$; $y_m = 8$

2.

x	2	8	12	4	14	6	16	10
y	6	11	16	7	17	10	19	14

$x_m = 9$; $y_m = 12.5$

3. Plot the data points for Exercise 1.

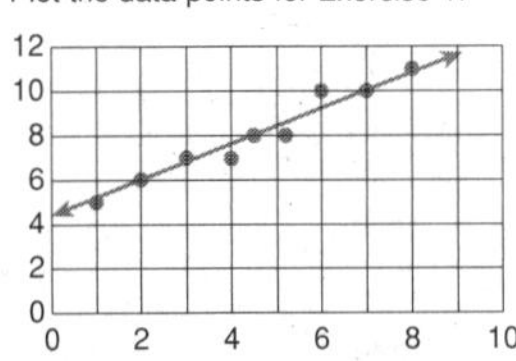

4. Plot the data points for Exercise 2.

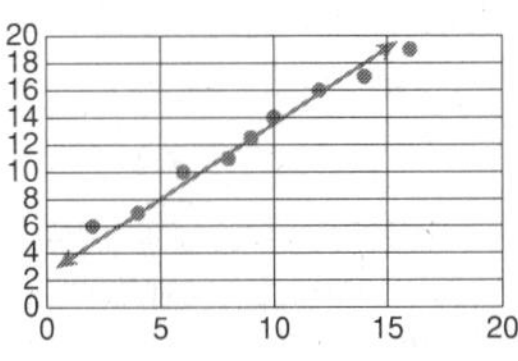

5. Draw a line through (x_m, y_m) of Exercise 3 to best fit the data.

Possible answer is shown above.

6. Draw a line through (x_m, y_m) of Exercise 4 to best fit the data.

Possible answer is shown above.

7. Write the equation of the line of best fit for Exercise 5.

$y = 0.8x + 4.4$

8. Write the equation of the line of best fit for Exercise 6.

$y = 1.1x + 2.6$

LESSON **12-7**

Practice B

Lines of Best Fit

Plot the data and find a line of best fit.

1.

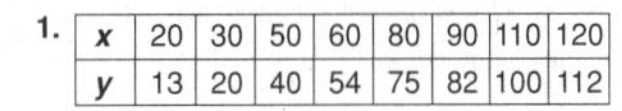

x	20	30	50	60	80	90	110	120
y	13	20	40	54	75	82	100	112

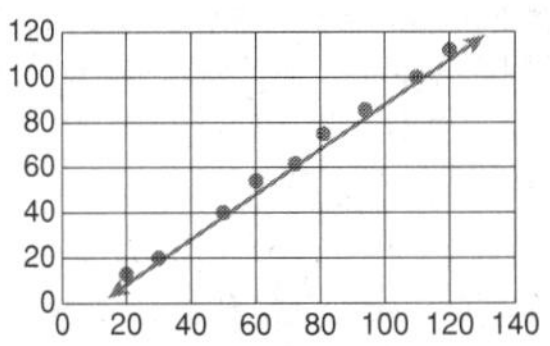

$y = x - 8$

2.

x	1.9	2.9	4.8	2.5	3.9	2.3	6.3	3.4
y	26	34	58	31	52	27	76	48

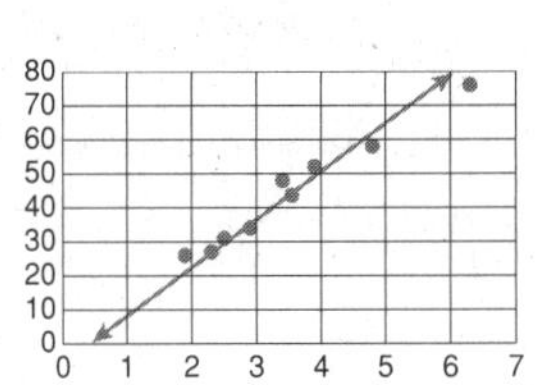

$y = 14.2x - 5.7$

3. Find the line of best fit for the student enrollment data. Use the equation of the line to predict what the enrollment at Columbus Junior High School will be in year 10. Is it reasonable to make this prediction? Explain.

Enrollment	405	485	557	593	638	712
Year	1	2	3	4	5	6

$x_m = 3.5$; $y_m = 565$; Possible equation of line of best fit: $y = 66x + 334$; Possible enrollment prediction for year 10: 994 students, which would be reasonable if the communities surrounding the school continued to grow at the same pace

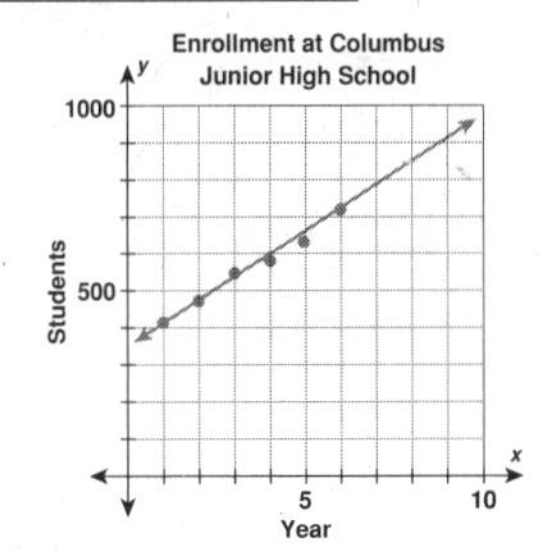

LESSON **12-7**

Practice C

Lines of Best Fit

Tell whether a line of best fit for each scatter plot would have a positive or negative slope. If a line of best fit would not be appropriate for the data, write *neither*.

1.

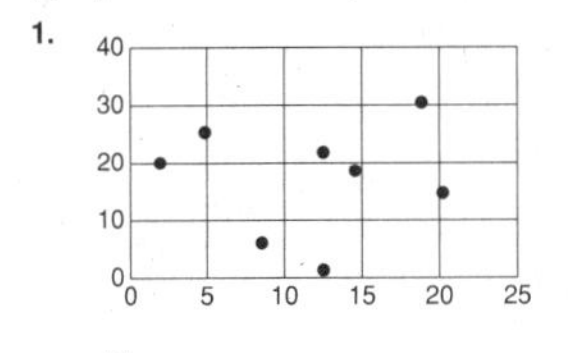

neither

2.

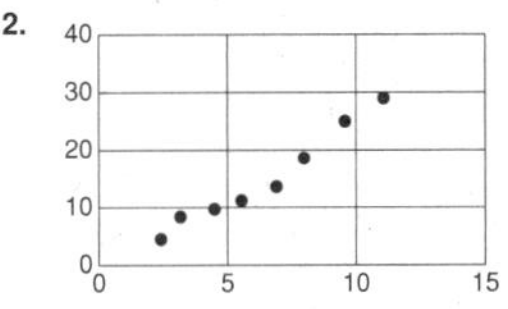

positive

3.

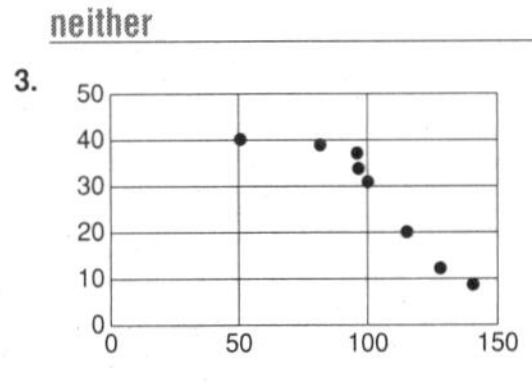

negative

4.

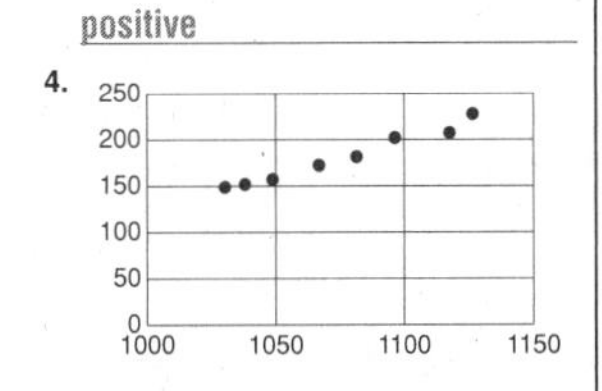

positive

5. Mrs. Kani is a salesperson for a local manufacturer. She wants to ask for a raise based on her sales record for the last eight months. Find the line of best fit for her sales data and use the equation of the line to predict her sales for one month a year from now.

Sales ($)	7500	7740	7790	7900	8130	8200	8300	8480
Month	1	2	3	4	5	6	7	8

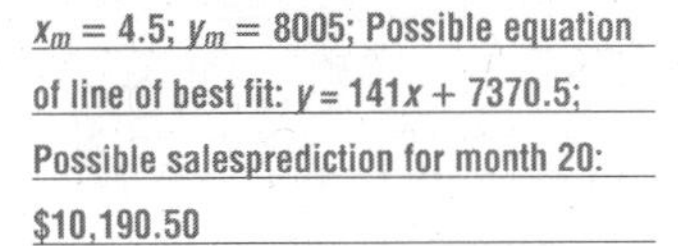

$x_m = 4.5$; $y_m = 8005$; Possible equation of line of best fit: $y = 141x + 7370.5$; Possible salesprediction for month 20: $10,190.50

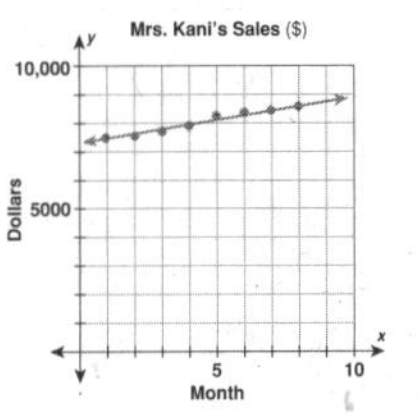

LESSON **12-7**

Reteach

Lines of Best Fit

Line of best fit: a line drawn near the points on a scatter plot to show the trend between two sets of data

To draw a line of best fit:

Draw the line through as many points as you can.
Try to get an equal number of points above the line as below.
Ignore any outliers.
It may happen that none of the points lie on the line.

The line of best fit for the data in the scatter plot below slants up, indicating a positive correlation.

The slope of this line of best fit is positive and its *y*-intercept is about 3.

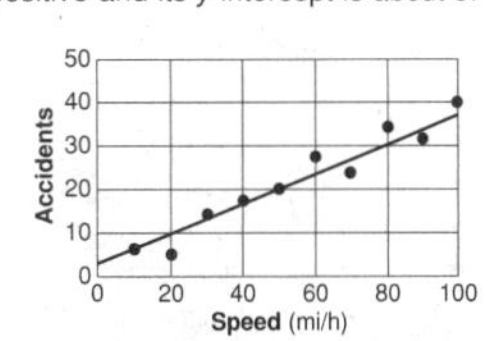

The line of best fit for the data in the scatter plot below slants down, indicating a negative correlation.

The slope of this line of best fit is positive and its *y*-intercept is about 28.

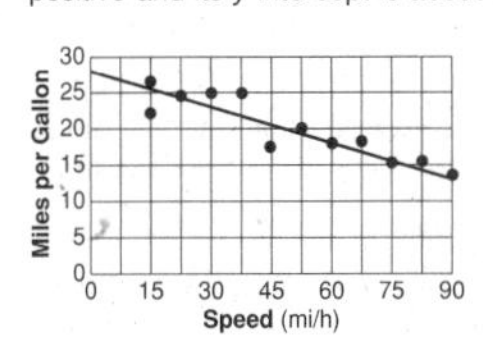

Draw a line of best fit for each graph. Describe the slope and find an approximate value for the *y*-intercept of the line of best fit. Possible answers:

1.

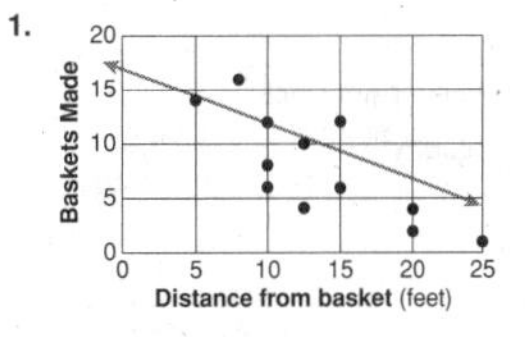

The line of best fit slants down.

The slope of the line is negative.

The *y*-intercept is approximately 17.

2.

The line of best fit slants up.

The slope of the line is positive.

The *y*-intercept is approximately 53.

LESSON 12-7

Reteach
Lines of Best Fit (continued)

To write an equation for a line of best fit, you can use the slope-intercept form, $y = mx + b$.

After you have drawn the line of best fit, estimate its slope from any two points on the line whose coordinates you can read.

Draw a right triangle with the hypotenuse on the line of best fit.

$$\text{slope} = \frac{\text{length of vertical leg}}{\text{length of horizontal leg}}$$

$\text{slope} = -\frac{2}{1}$, *y*-intercept = 10

equation for the line of best fit:
$y = -2x + 10$

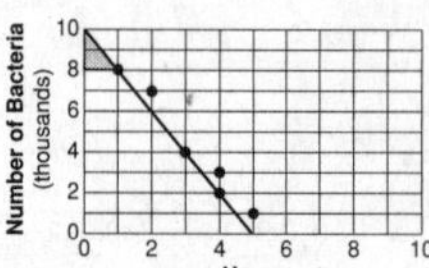

The data shows the number of bacteria present in a culture that has been treated with an anti-bacterial. The line of best fit can be used to predict the number of bacteria present after 2.5 hours.

$y = -2x + 10$
$y = -2(2.5) + 10$ Substitute $x = 2.5$.
$= -5 + 10$ Solve for *y*.
$= 5$
So, after 2.5 hours, the expected number of bacteria in the culture is approximately 5,000.

Find an equation for each line of best fit. Use the equation to answer the question.

3.

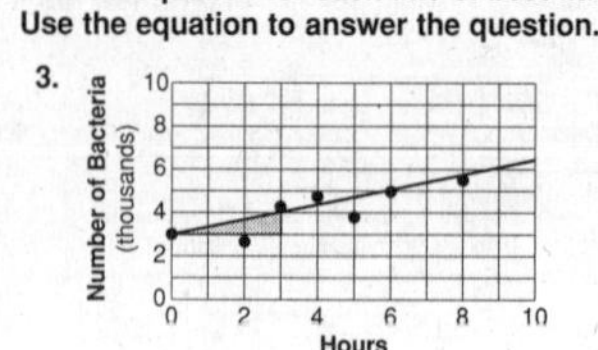

$m = \frac{1}{3}$, $b = 3$,

equation: $y = \frac{1}{3}x + 3$

The expected number of bacteria after 10 hours is about 6,333.

4.

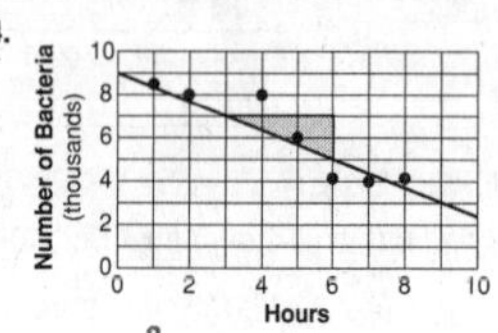

$m = -\frac{2}{3}$, $b = 9$,

equation: $y = -\frac{2}{3}x + 9$

The expected number of bacteria after 10 hours is about 2,333.

LESSON 12-7

Challenge
Use the Power of Technology

You can use a calculator to write an equation for the line of best fit. The following instructions are for the TI-83.

x	2	4	5	1	3	8	6	7
y	4	8	7	3	4	8	5	9

To enter the data in the calculator:

Display the Statistics menu.	Press STAT
Choose the EDIT option.	Press ENTER
Enter the *x*-values into list L1.	Press 2 ENTER Press 4 ENTER
Move to list L2.	Press ► Clear if necessary.
Enter the *y*-values into list L2.	Press 4 ENTER Press 8 ENTER

To get the slope and *y*-intercept for the line of best fit for this data set:

Display the Y = editor.	Press Y =
Display the Statistics menu.	Press STAT
Select the 4th option from CALC.	Press ► 4
Your screen reads **LinReg(*ax* + *b*)**.	
Attach L1 and L2 to your screen.	Press 2nd L1 , 2nd L2 ,
Attach Y1 to your screen.	Press VARS ► ENTER 1
Get the slope and *y*-intercept.	Press ENTER
The top part of your screen reads	**LinReg**
	y* = *ax* + *b
	***a* = .7380952381** This is the slope.
	***b* = 2.678571429** This is the *y*-intercept.

Use the information to write an equation for the line of best fit. $y = 0.74x + 2.68$
Compare these values to those in your text for Example 1.

Use a calculator to write an equation for the line of best fit for this data set.

x	3	8	4	4	5	7	1	6
y	5	14	8	9	12	3	2	11

$y = 1.13x + 2.65$

LESSON 12-7

Problem Solving
Lines of Best Fit

Write the correct answer. Round to the nearest hundredth.

1. The table shows in what year different average speed barriers were broken at the Indianapolis 500. If *x* is the year, with $x = 0$ representing 1900, and *y* is the average speed, find the mean of the *x*- and *y*-coordinates.

   $x_m = 52$; $y_m = 132.47$

Barrier (mi/h)	Year	Average Speed (mi/h)
80	1914	82.5
100	1925	101.1
120	1949	121.3
140	1962	140.9
160	1972	163.0
180	1990	186.0

2. Graph the data from exercise 1 and find the equation of the line of best fit.

   Possible answer:

   $y = 1.34x + 62.79$

3. Use your equation to predict the year the 210 mph barrier will be broken.

   Possible answer: 2010

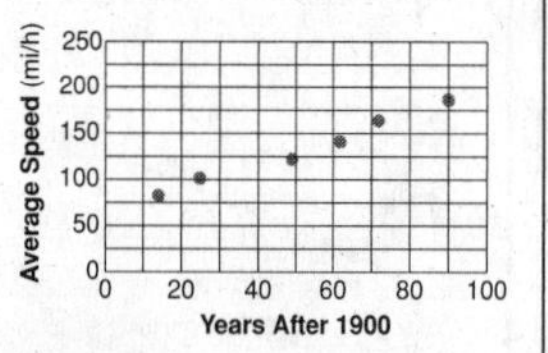

The percent of the U.S. population who smokes can be represented by the line of best fit with the equation $y = -0.57x + 44.51$ where *x* is the year, $x = 0$ represents 1960, and *y* is the percent of the population who smokes. Circle the letter of the correct answer.

4. Which term describes the percent of the population that smokes?
   A Increasing C No change
   (B) Decreasing D Cannot tell

5. Use the equation to predict the percent of smokers in 2005.
   F 13.16% H 24.56%
   (G) 18.86% J 41.66%

6. Use the equation to predict when the percent of smokers will be less than 15%.
   A 1996 (C) 2012
   B 2010 D 2023

7. Use the equation to predict the percent of smokers in 2010.
   F 10.31% (H) 16.01%
   G 11.01% J 12.71%

LESSON 12-7

Reading Strategies
Focus on Vocabulary

You may have T-shirts in different sizes, but none that really fits you. Your mom may tell you, "Choose the one that fits best." She means the shirt that comes the closest to fitting you.

When the points on a scatter plot do not lie in a straight line, you can draw a **line of best fit.**

Draw a line of best fit so that there are about the same number of points above the line as below the line.

The points on this scatter plot show the number of sit-ups Tim completes each minute. The line of best fit shows **about** how many sit-ups Tim completes each minute.

Tim's Sit-ups

Use the scatter plot to answer the following questions.

1. What information is shown along the bottom edge of the scatter plot?

   number of minutes

2. What information is shown along the left edge of the scatter plot?

   number of sit-ups

3. Do the points on a scatter plot lie exactly on a straight line?

   no

4. Does the line of best-fit show actual data?

   no

5. Using the line of best fit, predict how many sit-ups Tim might complete in six minutes?

   100

LESSON **12-7**

Puzzles, Twisters & Teasers

Ups and Downs!

Draw a line of best fit for each scatter plot. Select the answer that shows how many points fall above the line of best fit. To solve the riddle, match the letters of the answers with the blanks in the riddle.

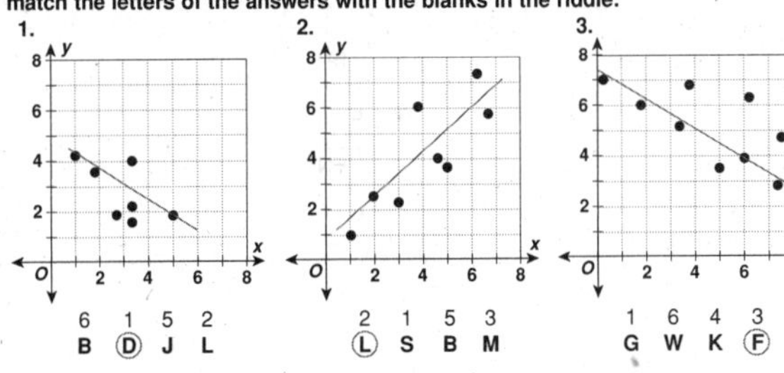

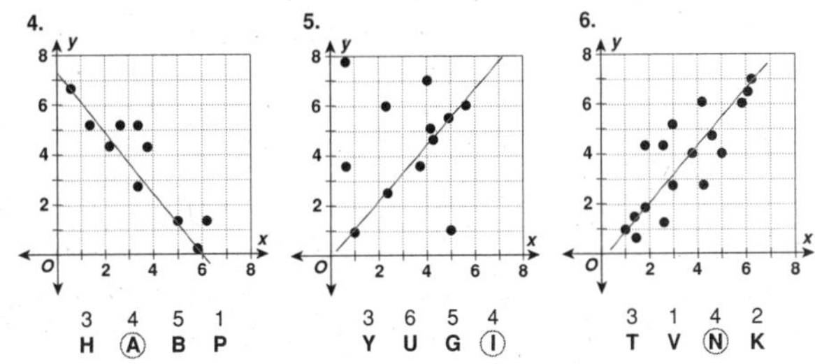

Where do fish like to go on vacation?

F I N L A N D
3 6 5 2 4 5 1